MW01234223

Scott, Foresman

Discover SCIENCE

Authors

Dr. Michael R. Cohen
Professor of Science and Environmental Education
School of Education
Indiana University
Indianapolis, Indiana

Dr. Timothy M. Cooney
Professor of Earth Science and Science Education
Earth Science Department
University of Northern Iowa
Cedar Falls, Iowa

Cheryl M. Hawthorne
Science Curriculum Specialist
Mathematics, Engineering, Science Achievement Program (MESA)
Stanford University
Stanford, California

Dr. Alan J. McCormack
Professor of Science Education
San Diego State University
San Diego, California

Dr. Jay M. Pasachoff
Director, Hopkins Observatory
Williams College
Williamstown, Massachusetts

Dr. Naomi Pasachoff
Research Associate
Williams College
Williamstown, Massachusetts

Karin L. Rhines
Science/Educational Consultant
Valhalla, New York

Dr. Irwin L. Slesnick
Professor of Biology
Western Washington University
Bellingham, Washington

Scott, Foresman and Company
Editorial Offices: Glenview, Illinois

Regional Offices: Sunnyvale, California • Tucker, Georgia •
Glenview, Illinois • Oakland, New Jersey • Dallas, Texas

Consultants

Special Content Consultant

Dr. Abraham S. Flexer
Science Education Consultant
Boulder, Colorado

Health Consultant

Dr. Julius B. Richmond
John D. MacArthur Professor of Health Policy
Director, Division of Health Policy Research and Education
Harvard University
Advisor on Child Health Policy
Children's Hospital of Boston
Boston, Massachusetts

Safety Consultant

Dr. Jack A. Gerlovich
Science Education Safety Consultant/Author
Des Moines, Iowa

Process Skills Consultant

Dr. Alfred DeVito
Professor Emeritus Science Education
Purdue University
West Lafayette, Indiana

Activity Consultants

Edward Al Pankow
Teacher
Petaluma City Schools
Petaluma, California

Valerie Pankow
Teacher and Writer
Petaluma City Schools
Petaluma, California

Science and Technology Consultant

Dr. David E. Newton
Adjunct Professor—Science and Social Issues
University of San Francisco
College of Professional Studies
San Francisco, California

Cooperative Learning Consultant

Dr. Robert E. Slavin
Director, Elementary School Program
Center for Research on Elementary and Middle Schools
Johns Hopkins University
Baltimore, Maryland

Gifted Education Consultants

Hilda P. Hobson
Teacher of the Gifted
W.B. Wicker School
Sanford, North Carolina

Christine Kuehn
Assistant Professor of Education
University of South Carolina
Columbia, South Carolina

Nancy Linkel York
Teacher of the Gifted
W.B. Wicker School
Sanford, North Carolina

Special Education Consultants

Susan E. Affleck
Classroom Teacher
Salt Creek Elementary School
Elk Grove Village, Illinois

Dr. Dale R. Jordan
Director
Jordan Diagnostic Center
Oklahoma City, Oklahoma

Dr. Shirley T. King
Learning Disabilities Teacher
Helfrich Park Middle School
Evansville, Indiana

Jeannie Rae McCoun
Learning Disabilities Teacher
Mary M. McClelland Elementary School
Indianapolis, Indiana

Thinking Skills Consultant

Dr. Joseph P. Riley II
Professor of Science Education
University of Georgia
Athens, Georgia

Reading Consultants

Patricia T. Hinske
Reading Specialist
Cardinal Stritch College
Milwaukee, Wisconsin

Dr. Robert A. Pavlik
Professor and Chairperson of Reading/Language Arts Department
Cardinal Stritch College

Dr. Alfredo Schifini
Reading Consultant
Downey, California

Cover painting commissioned by Scott, Foresman
Artist: Alex Gnidziejko

ISBN: 0-673-35685-X

23456789-RRW-0099989796959493

Reviewers and Content Specialists

Dr. Ramona J. Anshutz
Science Specialist
Kansas State Department of Education
Topeka, Kansas

Teresa M. Auldridge
Science Education Consultant
Amelia, Virginia

Annette M. Barzal
Classroom Teacher
Willetts Middle School
Brunswick, Ohio

James Haggard Brannon
Classroom Teacher
Ames Community Schools
Ames, Iowa

Priscilla L. Callison
Science Teacher
Topeka Adventure Center
Topeka, Kansas

Rochelle F. Cohen
Education Coordinator
Indianapolis Head Start
Indianapolis, Indiana

Linda Lewis Cundiff
Classroom Teacher
R. F. Bayless Elementary School
Lubbock, Texas

Dr. Patricia Dahl
Classroom Teacher
Bloomington Oak Grove Intermediate School
Bloomington, Minnesota

Audrey J. Dick
Supervisor, Elementary Education
Cincinnati Public Schools
Cincinnati, Ohio

Nancy B. Drabik
Reading Specialist
George Washington School
Wyckoff, New Jersey

Bennie Y. Fleming
Science Supervisor
Providence School District
Providence, Rhode Island

Mike Graf
Classroom Teacher
Branch Elementary School
Arroyo Grande, California

Thelma Robinson Graham
Classroom Teacher
Pearl Spann Elementary School
Jackson, Mississippi

Robert G. Guy
Classroom Teacher
Big Lake Elementary School
Sedro-Woolley, Washington

Dr. Claude A. Hanson
Science Supervisor
Boise Public Schools
Boise, Idaho

Dr. Jean D. Harlan
Psychologist, Early Childhood Consultant
Lighthouse Counseling Associates
Racine, Wisconsin

Dr. Rebecca P. Harlin
Assistant Professor of Reading
State University of New York—Geneseo
Geneseo, New York

Richard L. Ingraham
Professor of Biology
San José State University
San José, California

Ron Jones
Science Coordinator
Salem Keizer Public Schools
Salem, Oregon

Sara A. Jones
Classroom Teacher
Burroughs-Molette Elementary School
Brunswick, Georgia

Dr. Judy LaCavera
Director of Curriculum and Instruction
Learning Alternatives
Vienna, Ohio

Jack Laubisch
K-12 Science, Health, and Outdoor Education Coordinator
West Clermont Local School District
Amelia, Ohio

Douglas M. McPhee
Classroom Teacher/Consultant
Del Mar Hills Elementary School
Del Mar, California

Larry Miller
Classroom Teacher
Caldwell Elementary School
Caldwell, Kansas

Dr. Robert J. Miller
Professor of Science Education
Eastern Kentucky University
Richmond, Kentucky

Sam Murr
Teacher—Elementary Gifted Science
Academic Center for Enrichment—Mid Del Schools
Midwest City—Del City, Oklahoma

Janet Nakai
Classroom Teacher
Community Consolidated School District #65
Evanston, Illinois

Patricia Osborne
Classroom Teacher
Valley Heights Elementary School
Waterville, Kansas

Elisa Pinzón-Umaña
Classroom Teacher
Coronado Academy
Albuquerque, New Mexico

Dr. Jeanne Phillips
Director of Curriculum and Instruction
Meridian Municipal School District
Meridian, Mississippi

Maria Guadalupe Ramos
Classroom Teacher
Metz Elementary School
Austin, Texas

Elissa Richards
Math/Science Teacher Leader
Granite School District
Salt Lake City, Utah

Mary Jane Roscoe
Teacher and Team Coordinator
Fairwood Alternative Elementary School of Individually Guided Education
Columbus, Ohio

Sister Mary Christelle Sawicki, C. S. S. F.
Science Curriculum Coordinator
Department of Catholic Education Diocese of Buffalo
Buffalo, New York

Ray E. Smalley
Classroom Teacher/Science Specialist
Cleveland School of Science
Cleveland, Ohio

Anita Snell
Elementary Coordinator for Early Childhood Education
Spring Branch Independent School District
Houston, Texas

Norman Sperling
Chabot Observatory
Oakland, California

Sheri L. Thomas
Classroom Teacher
McLouth Unified School District #342
McLouth, Kansas

Lisa D. Torres
Science Coordinator
Lebanon School District
Lebanon, New Hampshire

Alice C. Webb
Early Childhood Resource Teacher
Primary Education Office
Rockledge, Florida

Tina Ziegler
Classroom Teacher
Evanston, Illinois

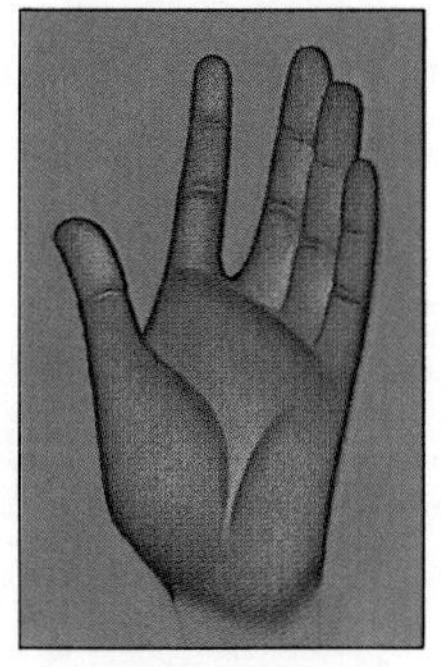
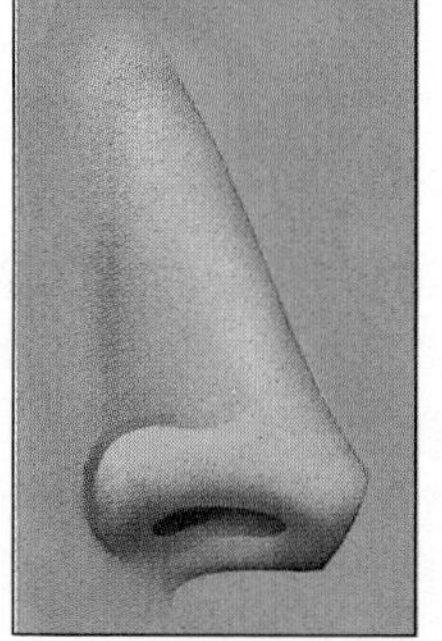
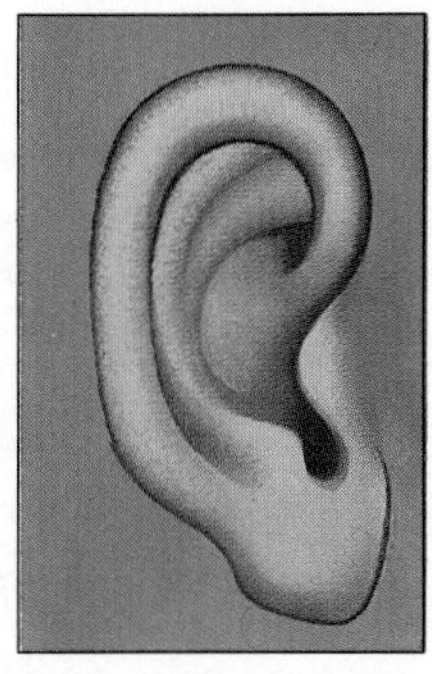
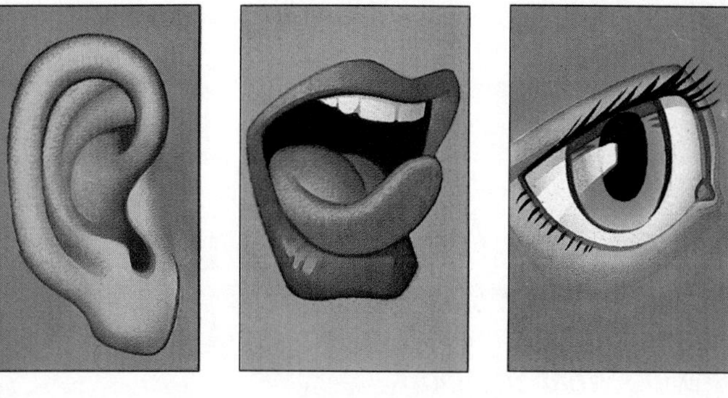
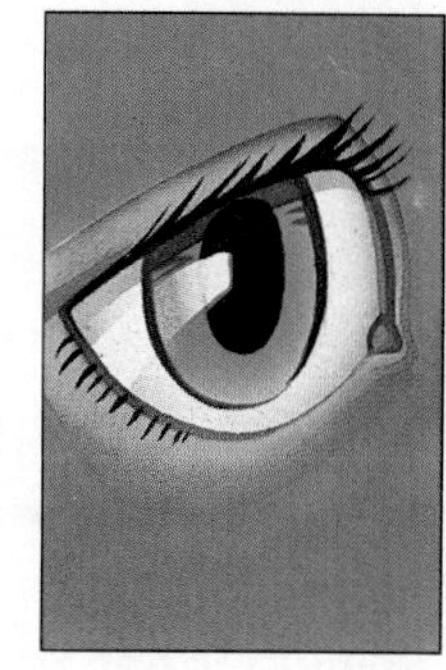

Chapter 2 Growing and Changing 30

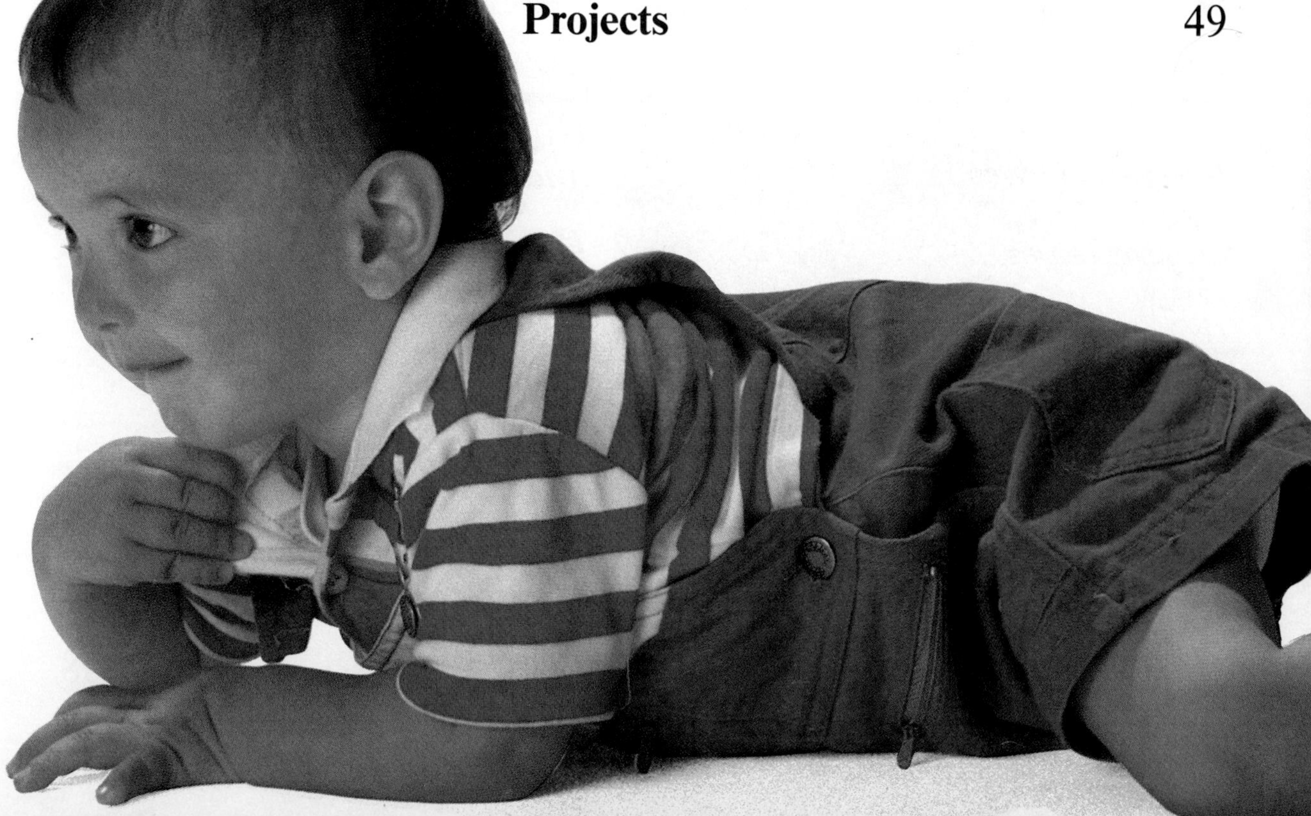

Unit 2 Life Science 50

Chapter 5 Learning About Animals 90

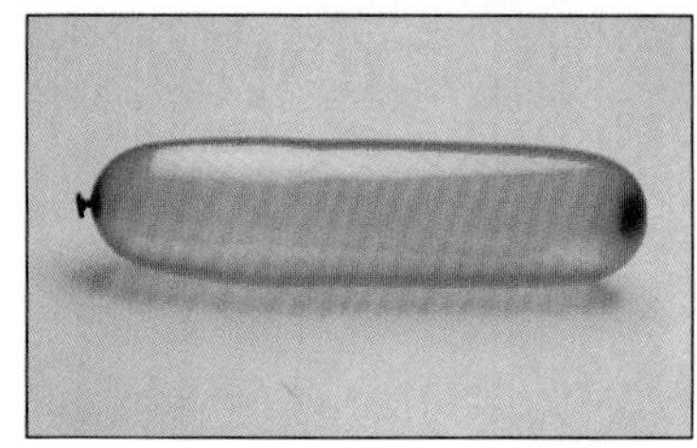

Chapter 8 Moving and Working 150

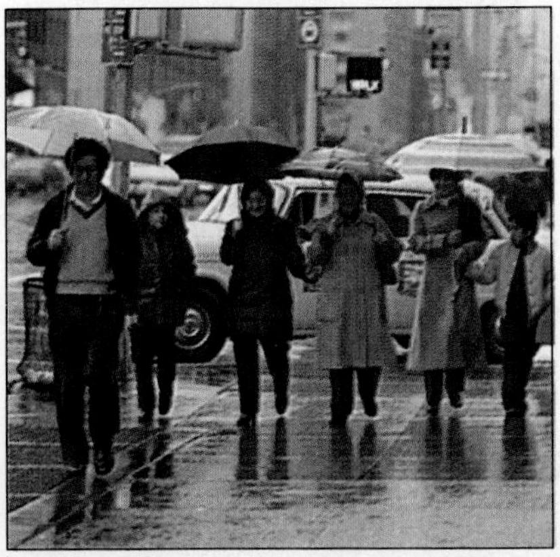

Discovering Science

Scientific Methods

Joe likes to learn about fish.
He is visiting the Aquarium with his class.
The Aquarium has many different kinds of fish.
Joe has never seen some of them before.

Joe was looking at some small fish.
Suddenly a big fish came right up to the glass.
Joe is surprised.
He thinks fish are fun to watch.

The fish are kept in large tanks.
Some tanks have fresh water in them.
Other tanks have salty water like the ocean.
Lights over the tanks help you see the fish.
Joe is looking at a saltwater tank.
Some of the animals look like plants.
These animals are corals.

Joe sees another large saltwater tank.
Many fish in the tank are very colorful.
Some of them have stripes or spots.
Some of the fish are hard to see.
Other ocean animals live in the tank.
A crab is in the corner of the tank.
Which animal looks like a flower?

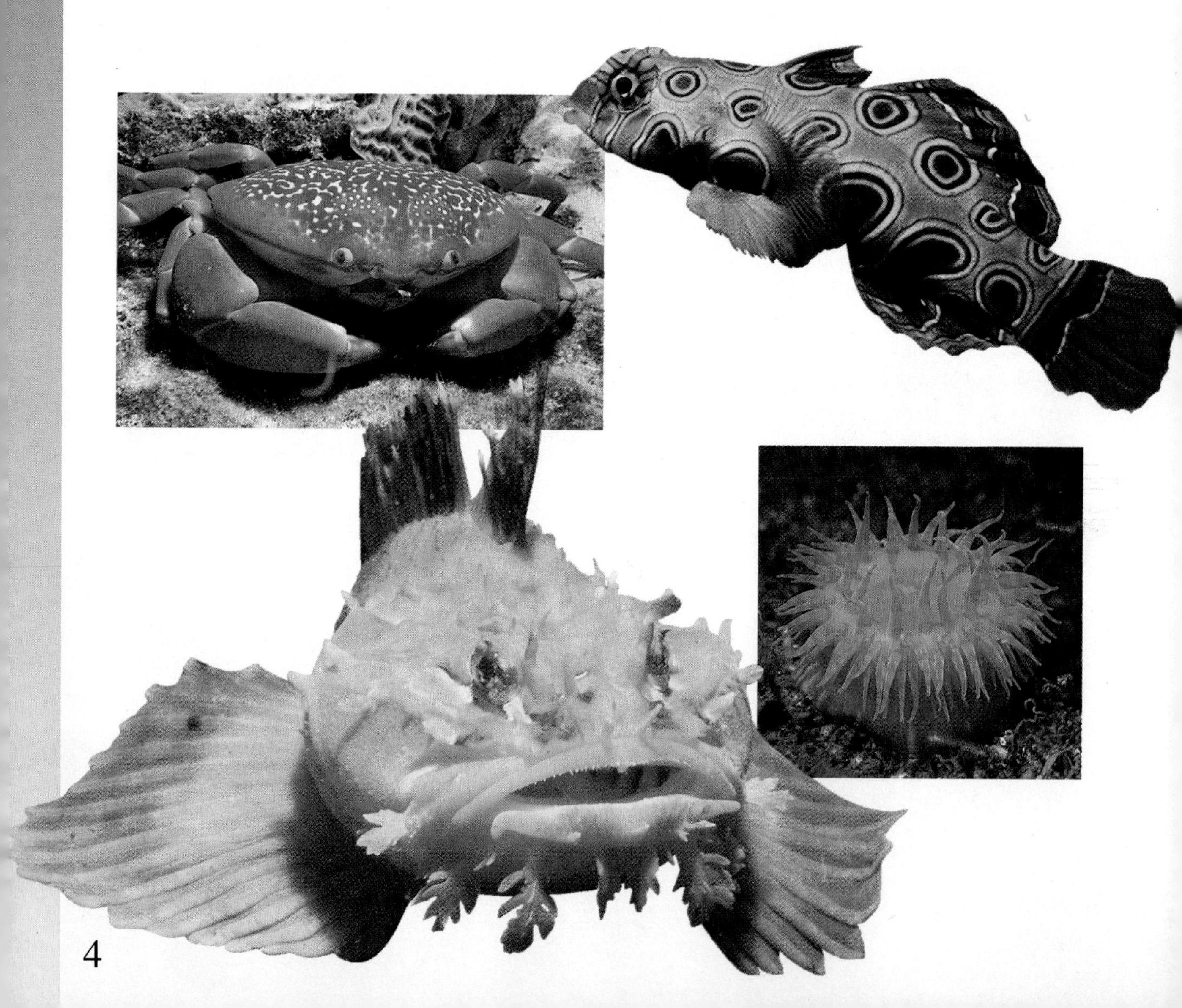

Some tanks have fresh water in them.
Freshwater fish live in rivers and lakes.
Joe notices the fish have different shapes.
Some fish are long and thin.
Some fish are flat.
Which animal has a hard covering?

Joe goes to another freshwater tank.
He sees some small colorful fish called tetras.
These fish are eating near the top of the tank.
Joe laughs at some fish with whiskers.
They are at the bottom of the tank.
These fish are called catfish.

Joe thinks about his catfish at home.
Do catfish always stay near the bottom?
Do other kinds of fish stay near the top?
Joe can use a scientific method to find out.

Scientists use scientific methods to find answers.
These methods have certain steps.
Read on to see how Joe uses the steps.
You can use these steps too.

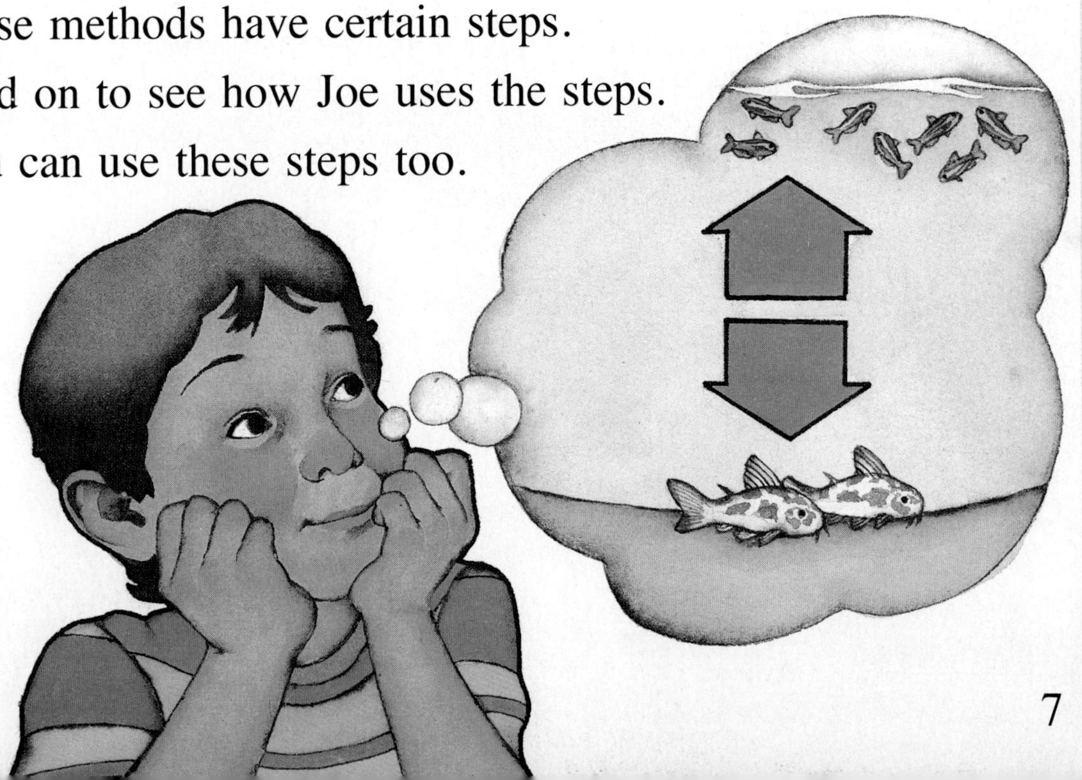

Explain the Problem

Joe has a question about fish.

Do fish live in certain places in the water?

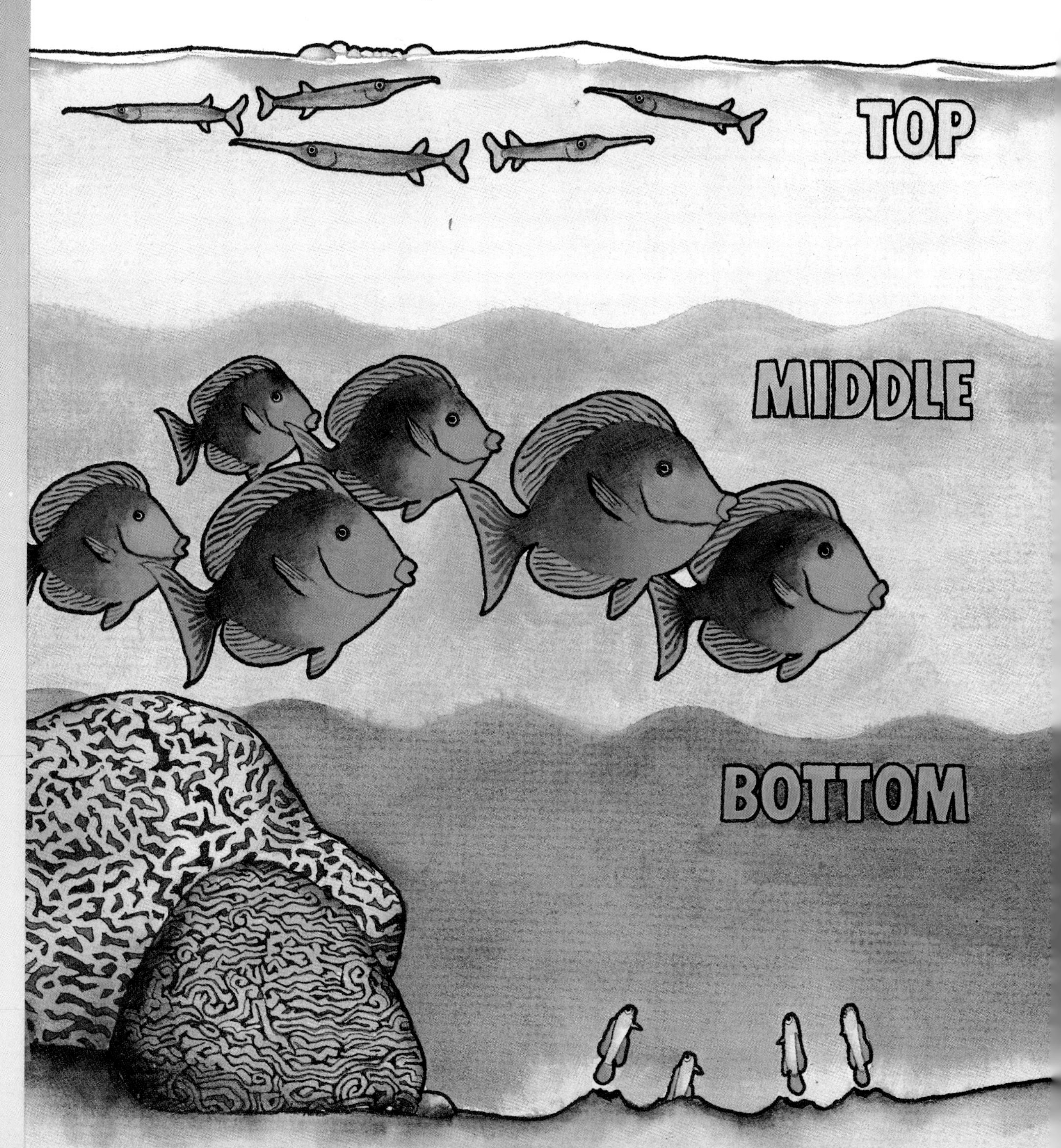

Make Observations

Joe saw many kinds of fish in a tank.

They had different colors and shapes.

He saw tetras near the top of the tank.

He noticed catfish on the bottom.

Give a Hypothesis

Joe thinks about his problem and observations.

He gives this hypothesis.

Fish live in certain places in the water.

Joe decides to do an experiment.

Joe uses the fish in his home aquarium.

He has five tetras and two catfish in it.

Make a Chart

Joe shakes some food on the water.
The tetras swim to the top of the tank to eat.
The catfish stay on the bottom.
They wait for the food to come down to them.

Joe looks at the tank two hours later.
The catfish are still near the bottom.
The tetras almost never go near the bottom.
Joe writes his observations in a chart.

Make Conclusions

Joe decides that his hypothesis is right.
Fish live in certain places in the water.
Notice Joe's conclusion.
How is it like his hypothesis?

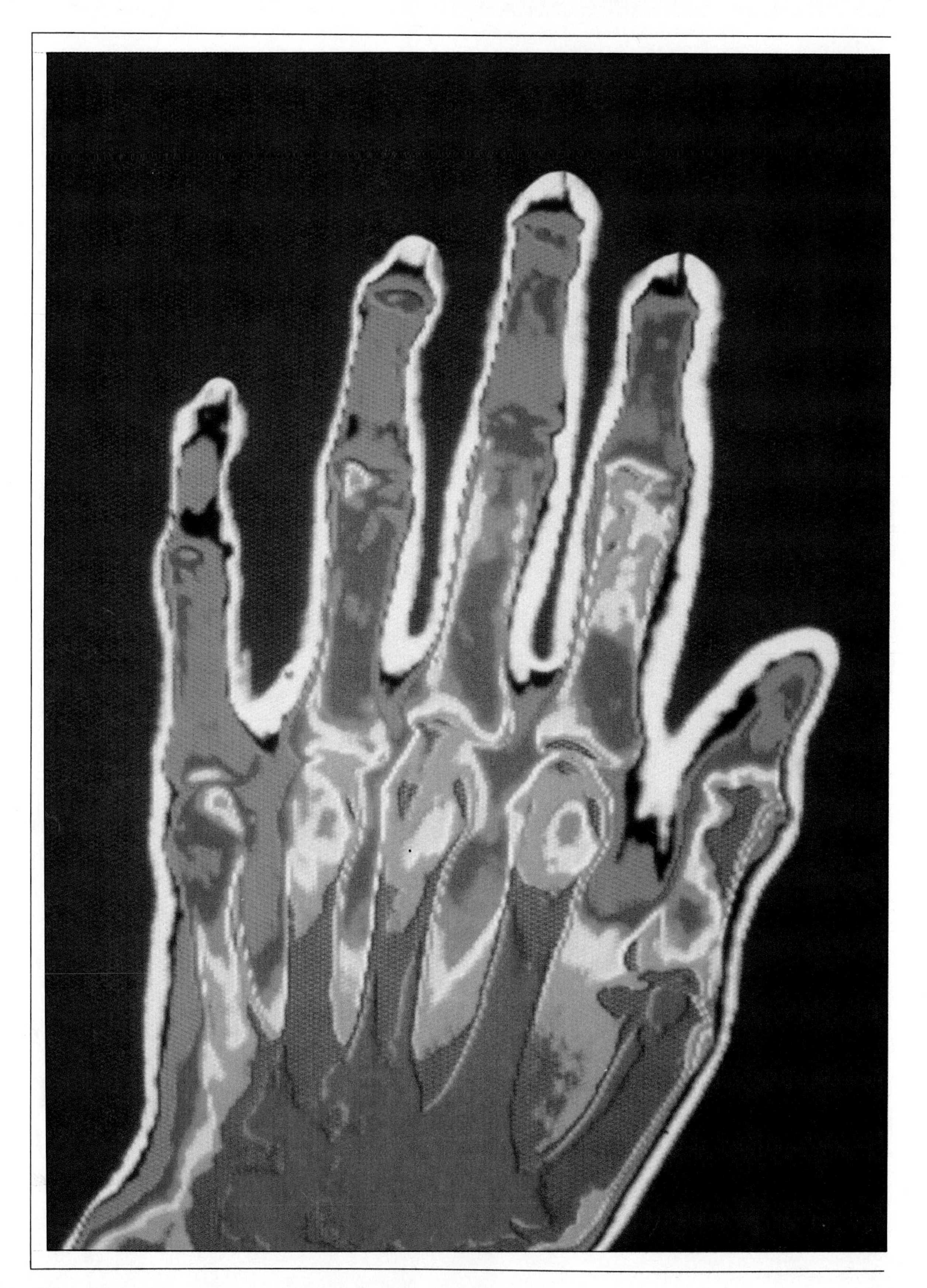

Unit 1

Human Body

Look at this special picture.
It shows a part of your body.
What part does the picture show?
Point to the bones in the picture.

Tell how you use this body part.
Then draw a picture of your own hand.

Chapter 1 Your Senses
Chapter 2 Growing and Changing

Chapter 1

Your Senses

What can you learn about water?
What parts of your body help you learn?

Starting the Chapter

How do you learn about things?

You can use body parts to learn.

Find out how body parts help you learn.

Then read more about learning.

TRY THIS

Learning About Things

Have someone cover your eyes.
Have the person give you things.
Feel, listen, and smell each thing.
What body parts do you use?
What did you learn?

Lesson 1 How Can You Learn?

You learn by using your **senses.**
Seeing and **hearing** are senses.
Touching and **tasting** are senses.
Smelling is another sense.
Pretend to use your senses.
What would you learn in this place?

What do you learn by using each sense?
You learn by feeling with your skin.
You learn by hearing with your ears.
You learn by seeing with your eyes.
What are these children learning?

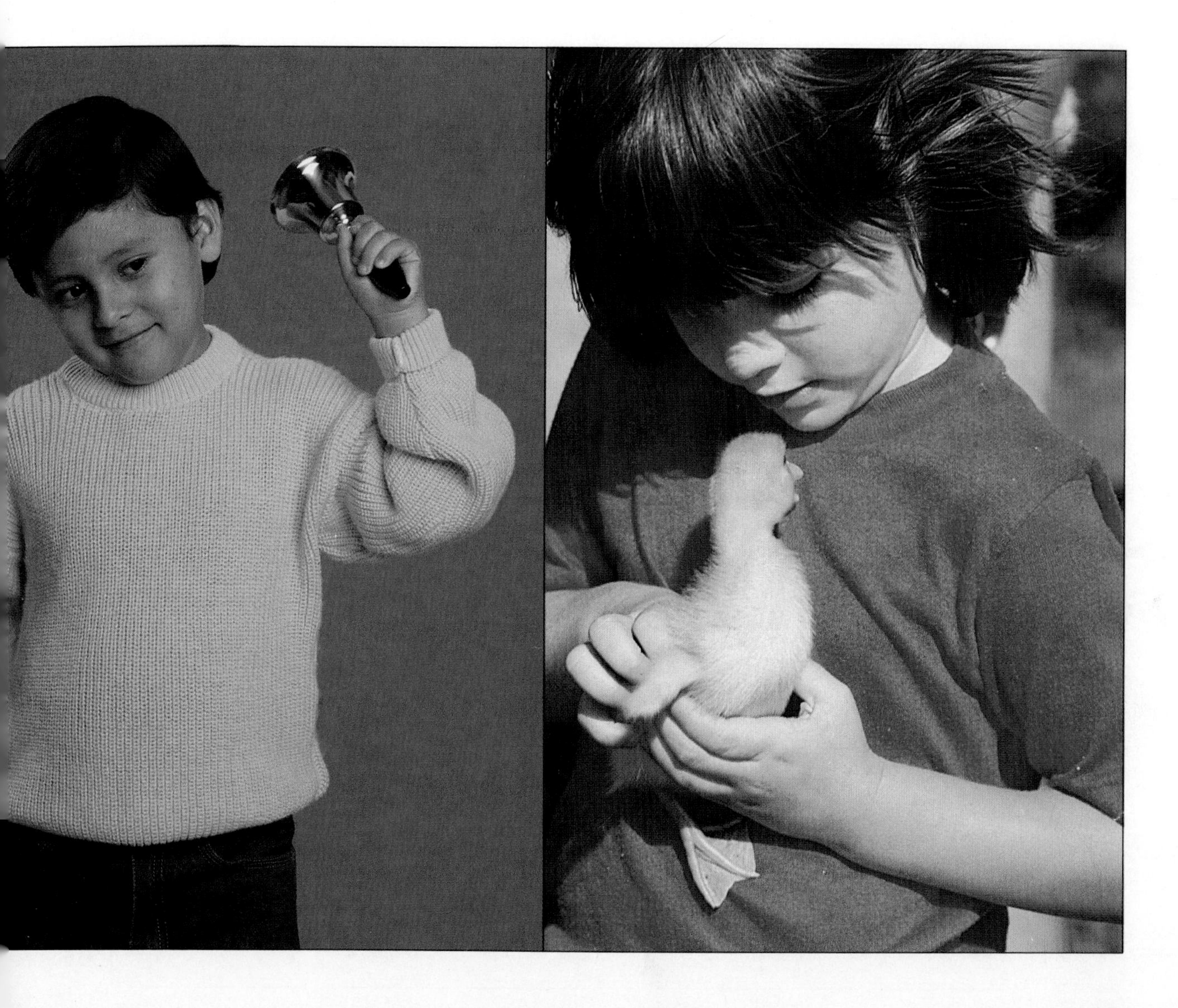

Your nose and tongue help you learn.
You use your nose to help you smell.
You use your tongue to help you taste.
What could you smell and taste here?

Lesson Review

1. What are your five senses?
2. How can your senses help you?

HUMAN BODY

Find Out

Close your eyes for a minute.
What did you hear?

ACTIVITY

Learning by Smelling

Follow the Directions

1. Pick up a cup and shake it gently.
2. Smell what is in the cup.
3. Tell what you think is in the cup.
4. Do the same for the other cups.

Tell What You Learned

Tell what you learned by smelling.
Draw three things you like to smell.

Lesson 2 How Do You Use Your Senses?

Your senses help you **observe.**

What do you do when you observe?

You notice many things.

You get information.

What could you observe here?

You use all your senses to observe.
You notice different sounds.
You notice different tastes.
How could you use your senses here?

Senses can help keep you safe.
You can hear warning sounds.
You can see colors and signs.
You can see cars and animals.
How do senses help with safety here?

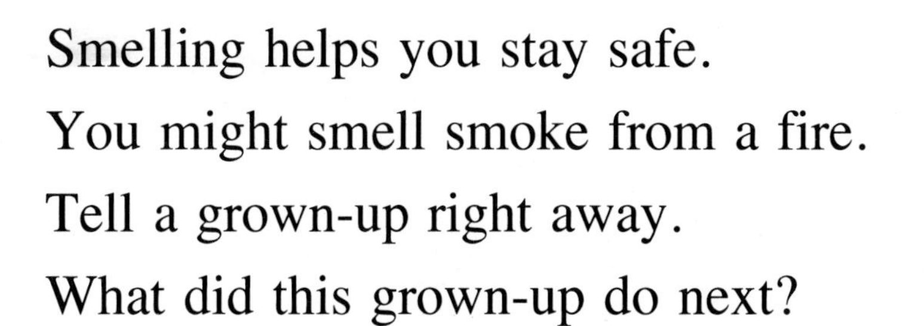

Smelling helps you stay safe.
You might smell smoke from a fire.
Tell a grown-up right away.
What did this grown-up do next?

Lesson Review

1. How do senses help you observe?
2. How can senses help with safety?

EARTH SCIENCE **Find Out** CONNECTION

Senses tell you about the weather.
Which senses give you information?
What words tell about weather?

Using All Your Senses

Follow the Directions

1. Look at a peanut in a shell.
2. Smell the shell and feel it.
3. Shake the shell. Listen.
4. Open the shell.
5. Look at the peanuts.

Tell What You Learned

Tell how you used each sense.

Tell what you found out.

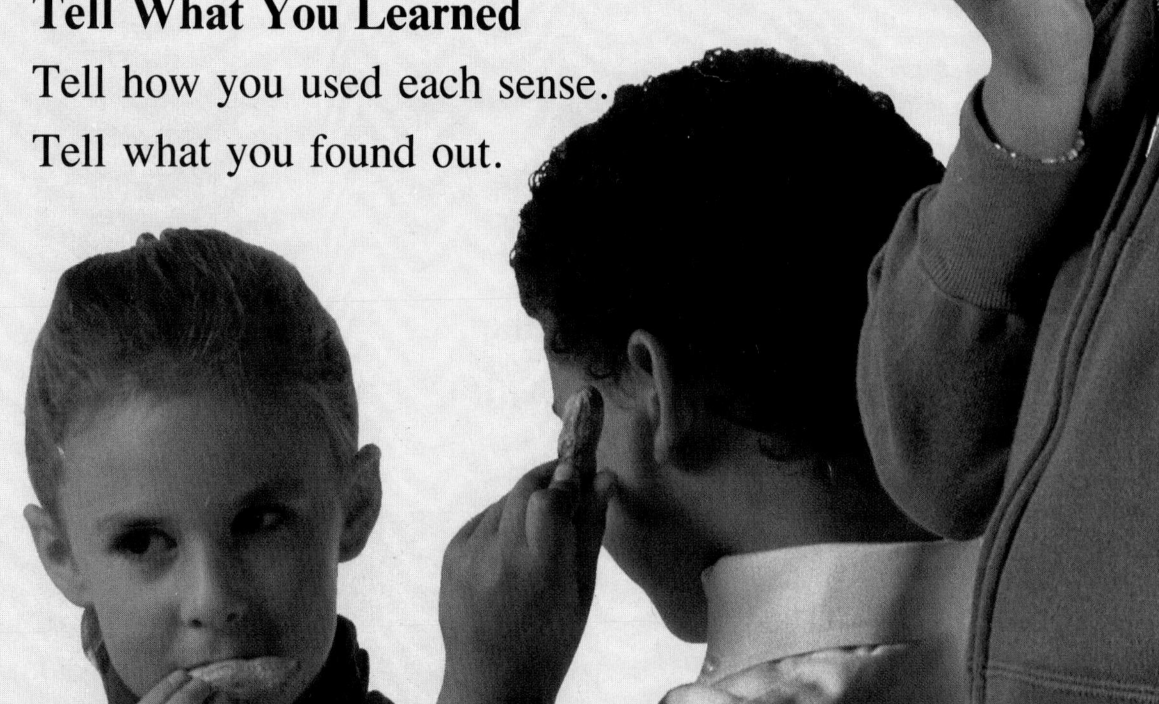

An Alarm to Help Smell Smoke

Fire can harm people.

Smoke from fires can warn people.

Suppose people do not smell the smoke.

Then this alarm can warn them.

It makes a noise when there is smoke.

Then people can go to a safe place.

What Do You Think?

Why would a person have this alarm?

Using a Pictograph

Which senses do you use to observe?

1. Look at the picture of the classroom.
   Find things to observe.
   Think of senses you would use.

2. Fill in the pictograph.
   Go across each row.
   Use small paper squares.
   Put one on each sense you would use.

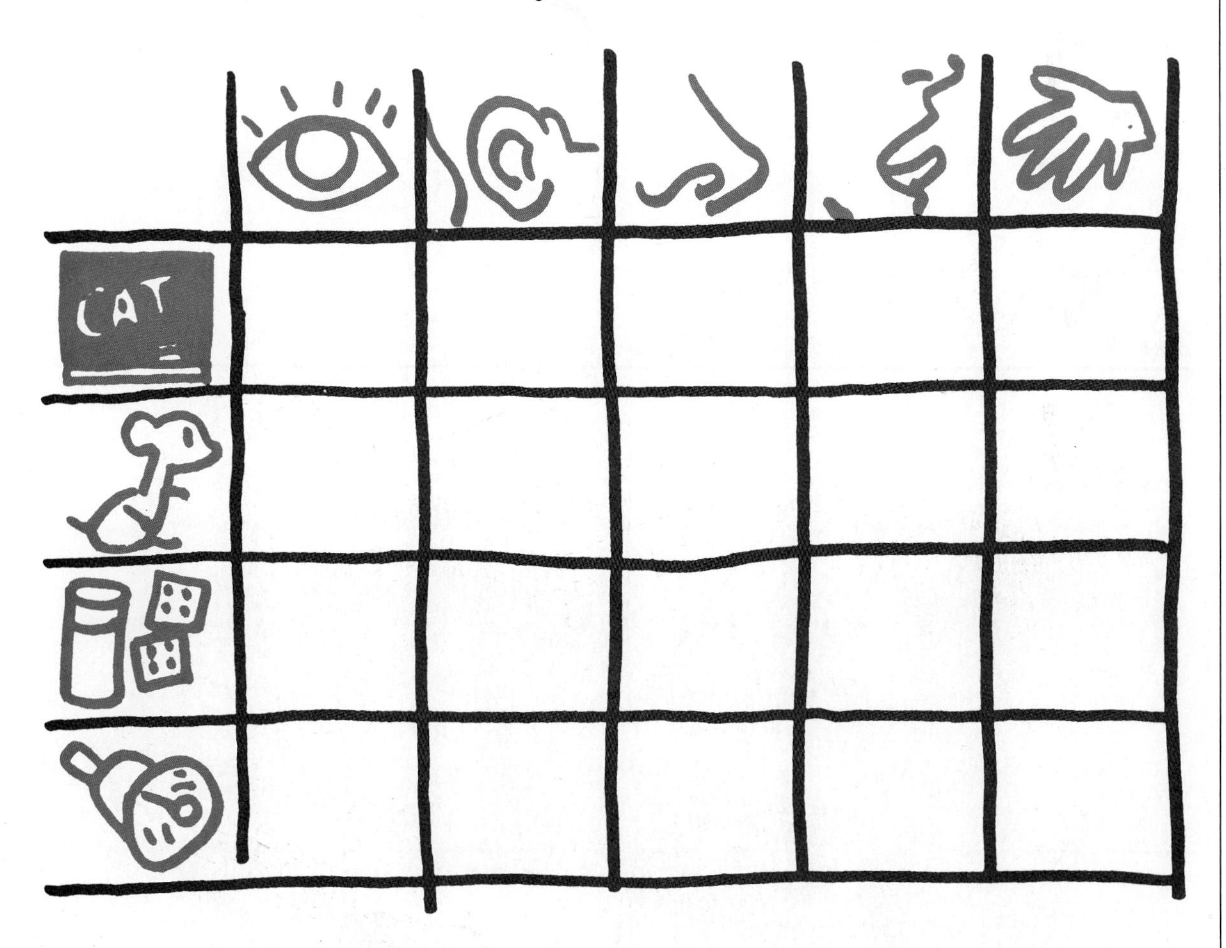

3. Look at your pictograph.
   Which sense did you use most?

Chapter 1 Review

Review Chapter Ideas

1. Name your senses.
2. Tell what senses help you learn.
3. Tell how senses help you observe.
4. Describe how senses help with safety.
5. Look at the pictures.

   Which senses give you warnings?

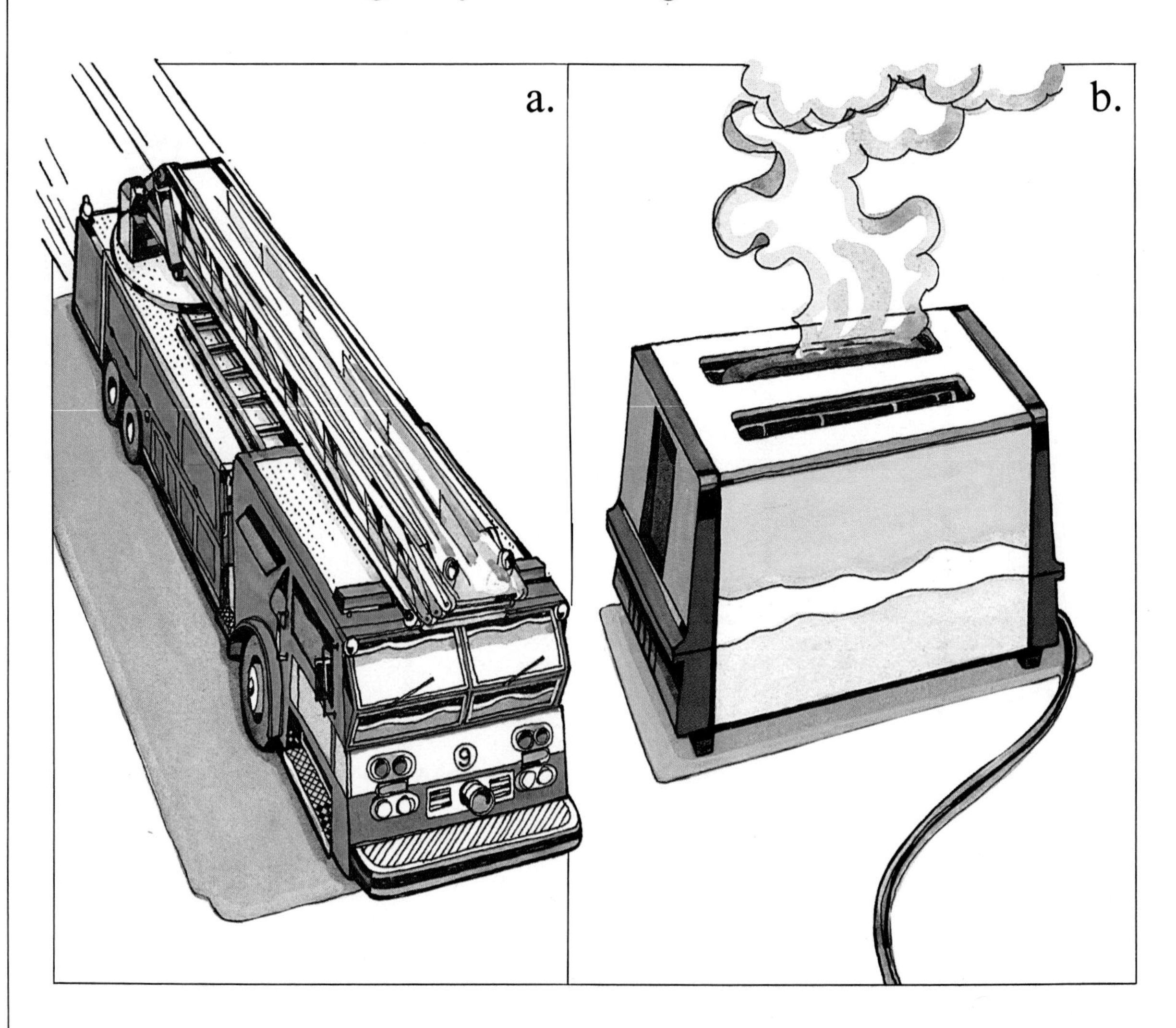

Review Science Words

Match the words and the pictures.

1. seeing
2. smelling
3. hearing
4. touching
5. tasting

a. b. c. d. e.

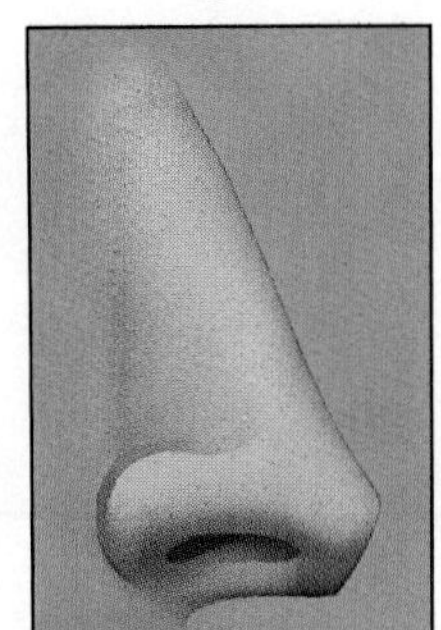

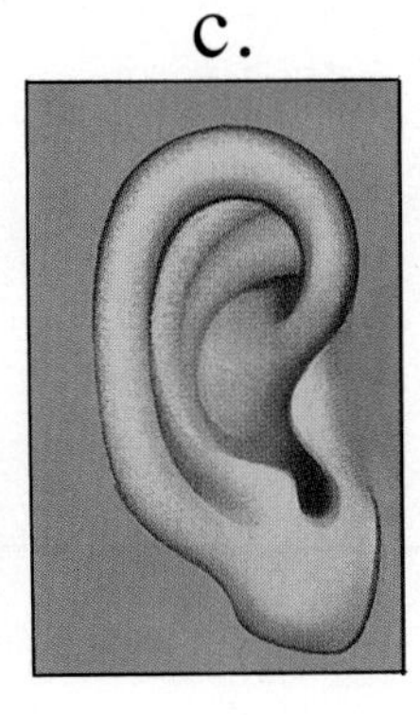

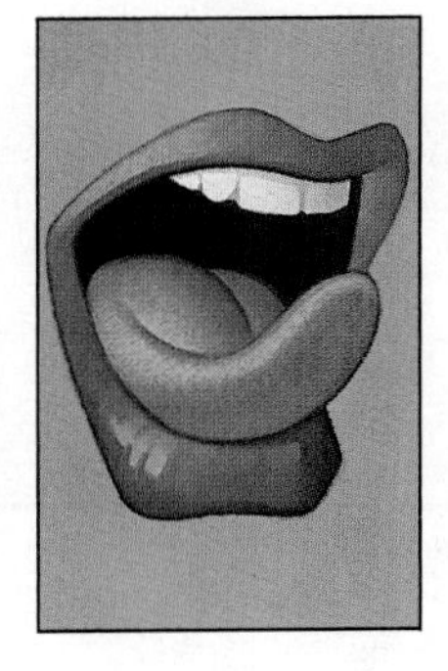

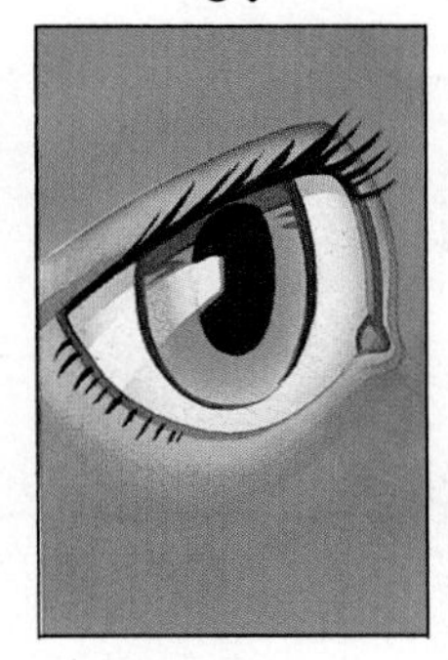

Tell what the words mean.

6. senses
7. observe

Use Science Ideas

What sense is being used for safety? How?

Chapter 2

Growing and Changing

How did this person grow and change?
How will you grow and change?

Starting the Chapter

Imagine your mother as a baby.

Could you pick out her picture?

Match pictures of your classmates.

Notice how they have changed.

Read more about changing.

TRY THIS

Matching Pictures

Bring your baby picture to school.
Ask your classmates to do the same.
Mix up the pictures.
Pick one and match it to a classmate.
Take turns matching the pictures.
What helps you match them?

Lesson 1 How Do People Change?

Your body changes as you grow.
Your size and shape change.
Your **bones** grow and you get taller.
Your **muscles** grow larger.
Who has grown more here?

Your teeth change as you grow.
Your first teeth are small.
Your jaws get larger as you grow.
Some of your teeth come out.
Then you get new, larger teeth.
The new teeth are **permanent teeth.**

What do you learn as you grow?
You learn new games and sports.
You learn new things in school.
You can learn more all your life.
What are these children learning?

You play differently as you grow.

How are these children playing?

You learn to get along with people.

Your family and friends help.

Lesson Review

1. How does your body grow and change?
2. What can you learn as you grow?

List ways you have changed.

Use words about learning.

HUMAN BODY

Find Out

Getting New Teeth

Follow the Directions

1. Look at your teeth.
2. Find and count any new teeth.
3. Ask your classmates to find and count their new teeth.
4. Ask when they got new teeth.
5. How many children have new teeth?

Tell What You Learned

Draw how your teeth look.

Tell when children get new teeth.

Daniel A. Collins

Dentists help us care for our teeth.

Daniel A. Collins is a dentist.

He has taught other dentists.

Some people had painful teeth and jaws.

Dr. Collins worked with other dentists.

They helped the people feel better.

What Do You Think?

Why do people study teeth?

Lesson 2 What Can Help You Grow?

Food helps you grow.
It helps you stay **healthy.**
You need food every day.
You need different kinds of food.
What foods do you see here?
Which of these foods do you like?

Exercise helps you grow.
Playing every day is good exercise.
Play the way these children do.
Run, jump, hop, and skip, too.
You also need rest to grow.
You need about ten hours of sleep.

Staying well helps you grow.
Washing helps you stay well.
This girl keeps her hair and body clean.

Lesson Review

1. What can help you grow?
2. How can keeping clean help you?

LIFE SCIENCE
Find Out
CONNECTION

Look at pictures of foods.
Do they come from animals or plants?
Make two lists.

Getting Clean

Wear cover goggles for this activity.

Follow the Directions

1. Find the best way to get clean.
2. Dip three fingers in flour.
3. Blow on the first finger and look.
4. Wipe the second finger and look.
5. Wash the third finger and look.

Tell What You Learned

Tell the best way to get clean.

Tell when to wash your hands.

Measuring and Making a Chart

What can you learn from measuring?

1. Use a ruler.

   Measure your hand.

   Ask two friends to measure their hands.

   Write the numbers on your paper.

2. Make a chart like the one shown.
   Use the numbers on your paper.
   Color in the chart.
   Show how long your hand is.
   Color in the chart for your friends.

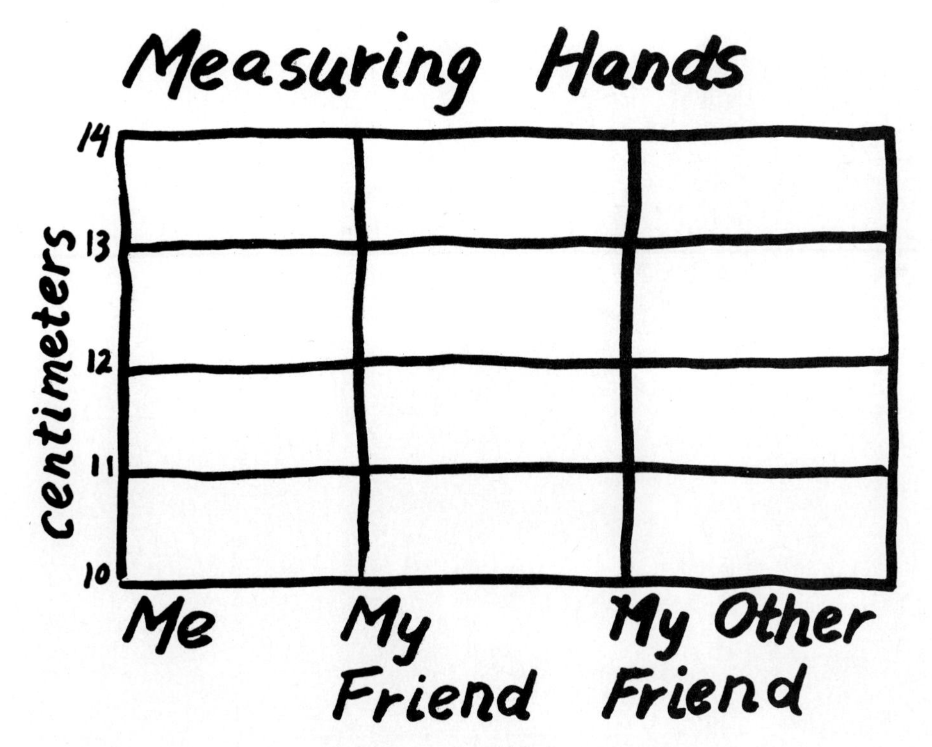

3. Look at the chart.
   What did measuring show?
   Does everyone grow the same?

Chapter 2 Review

Review Chapter Ideas

1. Tell how this baby will change.
2. Tell how teeth change as you grow.
3. Explain what you learn as you grow.
4. Tell what helps you grow.
5. Tell what good exercise is.
6. Tell how keeping clean helps you.

Review Science Words

Match the words and the pictures.

1. bones
2. exercise
3. muscles
4. permanent teeth

a. b. c. d.

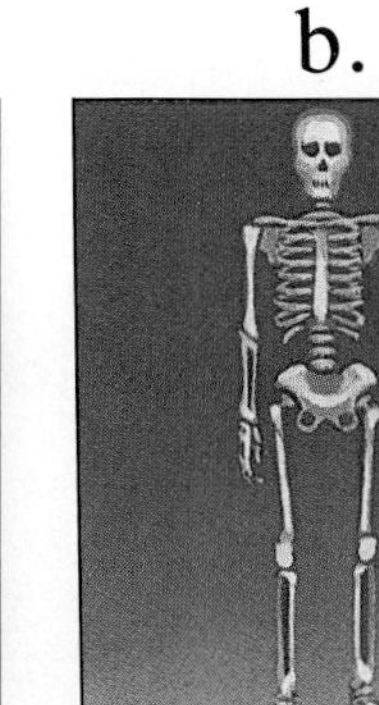

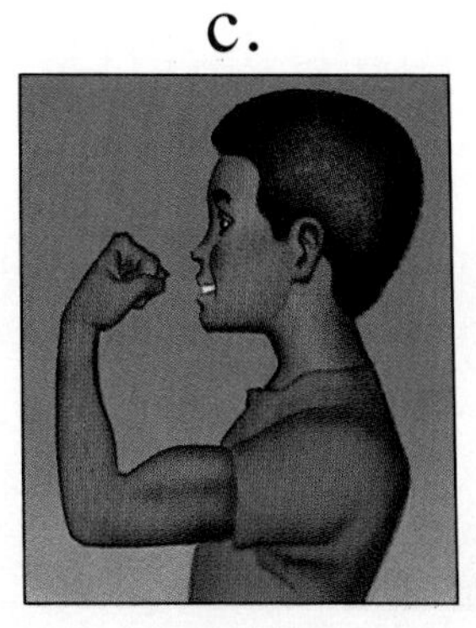

Tell what the word means.

5. healthy

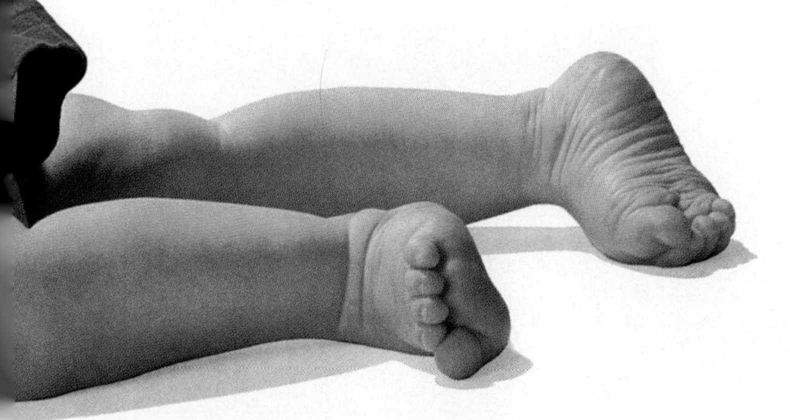

Use Science Ideas

What can help this baby stay healthy?

Dentists and Helpers

Dentists and their helpers study teeth.

They clean teeth and fix them.

They use special machines and tools.

They show people how to brush.

They show people how to floss.

Dentists help people care for teeth.

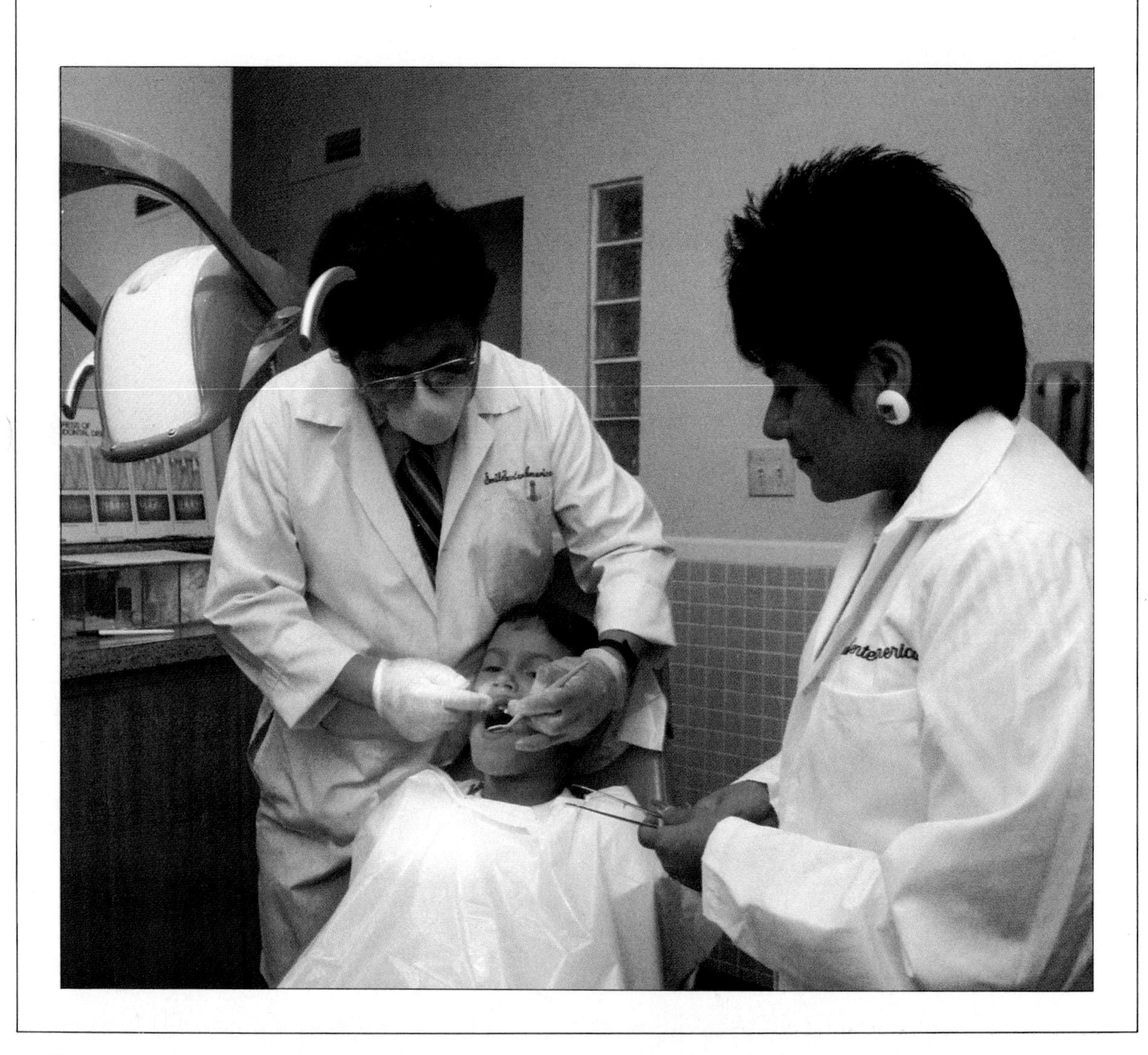

Stethoscope

This doctor uses a stethoscope.
It helps him hear body sounds.

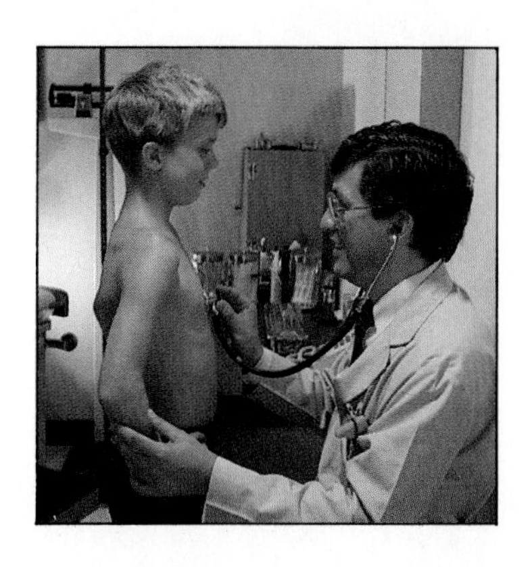

A stethoscope has two sides.
The smaller side is for low sounds.
The larger side is for higher sounds.

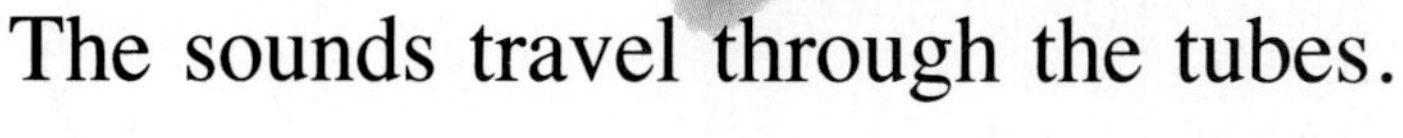

The sounds travel through the tubes.
Then the doctor can hear the sounds.

Unit 1 Review

Answer the Questions

1. How do you use your five senses?
2. What body parts help you learn?
3. How will you change as you grow?
4. Look at the picture.
   What changes in teeth do you see?
5. What helps you grow?

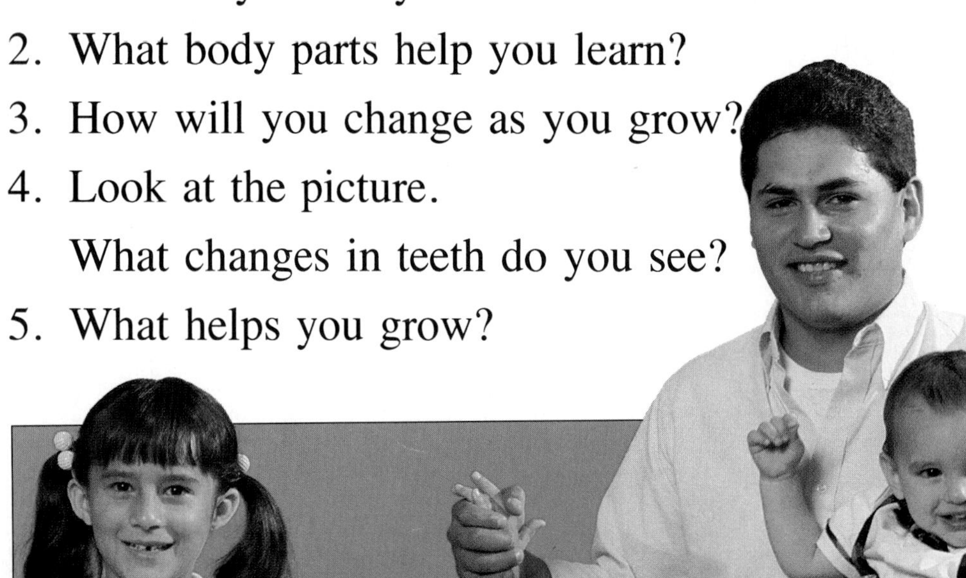

Study the Pictures

How could you use your senses here?

a.

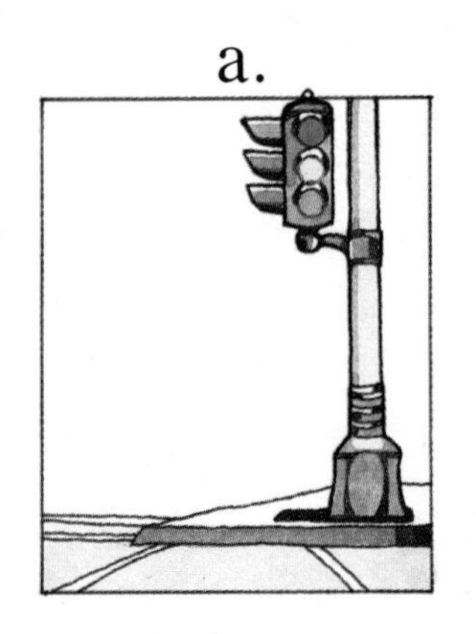

b.

c.

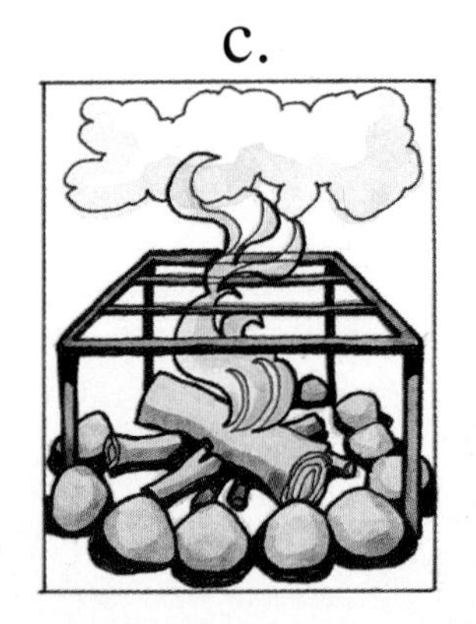

d.

What to Do

1. Look at the picture.
   Play the game.
   Ask a friend to put things in a box.
   Touch each thing and tell what it is.
   Put different things in the box.
   Ask a friend to guess.

2. Find some things to measure.
   Guess how many centimeters long each is.
   Then measure each thing.
3. Draw a picture of yourself.
   Show a way you help yourself grow.

Unit 2

Life Science

This frog and flower are living things.
They both need air to live.
The frog and flower need water, too.

What other living things can you name?
Work with someone in your class.
Think of other living things.

Chapter 3

Living and Nonliving

This figure looks alive.
What do you think it can do?
How can you tell it is not alive?

Starting the Chapter

You are alive.

You can do many things.

You can draw what you do.

Then read more about living things.

TRY THIS

Drawing What You Do

Draw three things you can do.
Look at the things in the picture.
Tell if they can do what you do.
What can living people do?

Lesson 1 What Is a Living Thing?

People are **living things.**
Animals and plants are living.
Many living things move on their own.
Living things can grow.
How will these animals grow?

Living things can change.
You can change and grow bigger.
Plants and animals grow bigger.
Some trees change and grow bigger.
How has this tree changed?

Where do new living things come from?
They come from other living things.
Living things can be **parents.**
Which are the new living things here?

Lesson Review

1. What are living things?
2. What can living things do?

LIFE SCIENCE
Find Out

Get pictures of living things.
Tell what the pictures show.

Observing Mealworms

Follow the Directions

1. Put a mealworm on your desk.
2. Put food near the mealworm.
3. Touch it and give it light.
4. Observe how the mealworm moves.

Tell What You Learned

Tell what your mealworm did.

Tell what another animal can do.

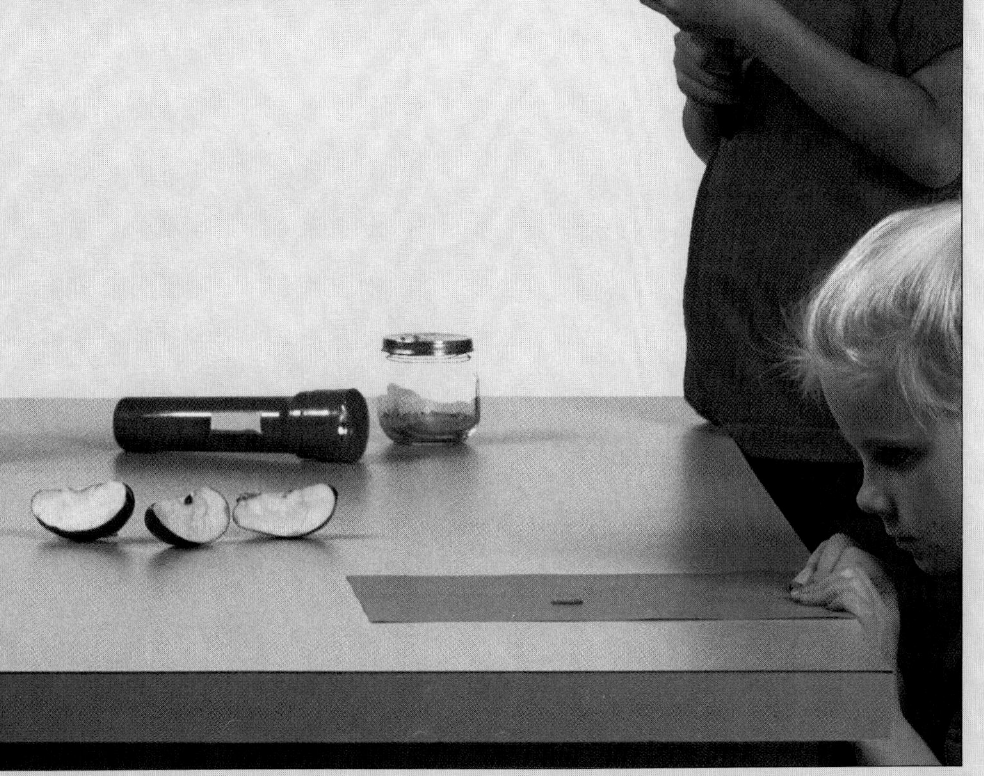

Lesson 2 What Do Living Things Need?

Living things need food.
Living things also need water.
Most living things need air.
What are these animals eating?
How do these plants get water?

Some living things need **shelter.**
A home gives shelter.
Some birds have homes in trees.
Some ants build homes in the ground.
What shelter does this animal have?

People need food.

They need air, water, and homes.

People also need to help each other.

How do these people help each other?

Lesson Review

1. What do plants and animals need?
2. What do people need?

LIFE SCIENCE

Find Out

Get pictures of animal homes.

Tell what the pictures show.

Science in Your Life

Traveling in Space

People travel in space.

Space has no air, water, or food.

People travel in spacecraft.

The craft have air, water, and food.

Then people can live and work in space.

What Do You Think?

What is shelter for people in space?

Lesson 3 What Are Nonliving Things Like?

Nonliving things were never alive.

They cannot do what living things do.

Nonliving things cannot eat food.

Can these things grow?

Are they living or nonliving?

Some nonliving things can move.
They do not move on their own.
Something makes nonliving things move.
What nonliving things do you see here?
Which ones can move?

Only living things can be parents.
Who are the living things here?
How do you know they are living?

Lesson Review

1. What does not eat, drink, or grow?
2. How do nonliving things move?

PHYSICAL SCIENCE
Find Out
CONNECTION

Find some nonliving things.
Find out which move.
Then tell what makes them move.

Discovering What Is Alive

Follow the Directions

1. Put some stones and beans on a tray.
2. Cover them with water.
3. Leave them alone for three or four days.
4. Tell what happens.

Tell What You Learned

Draw what was alive and not alive.

Draw another living thing.

Observing and Making a Graph

What can a graph show?

1. Observe the picture.

   Count the living things.

   Count the nonliving things.

   Write each number on your paper.

2. Copy the graph you see.
   Use the paper with the numbers you wrote.
   Use two colors.
   Color how many things are alive.
   Color how many things are not alive.

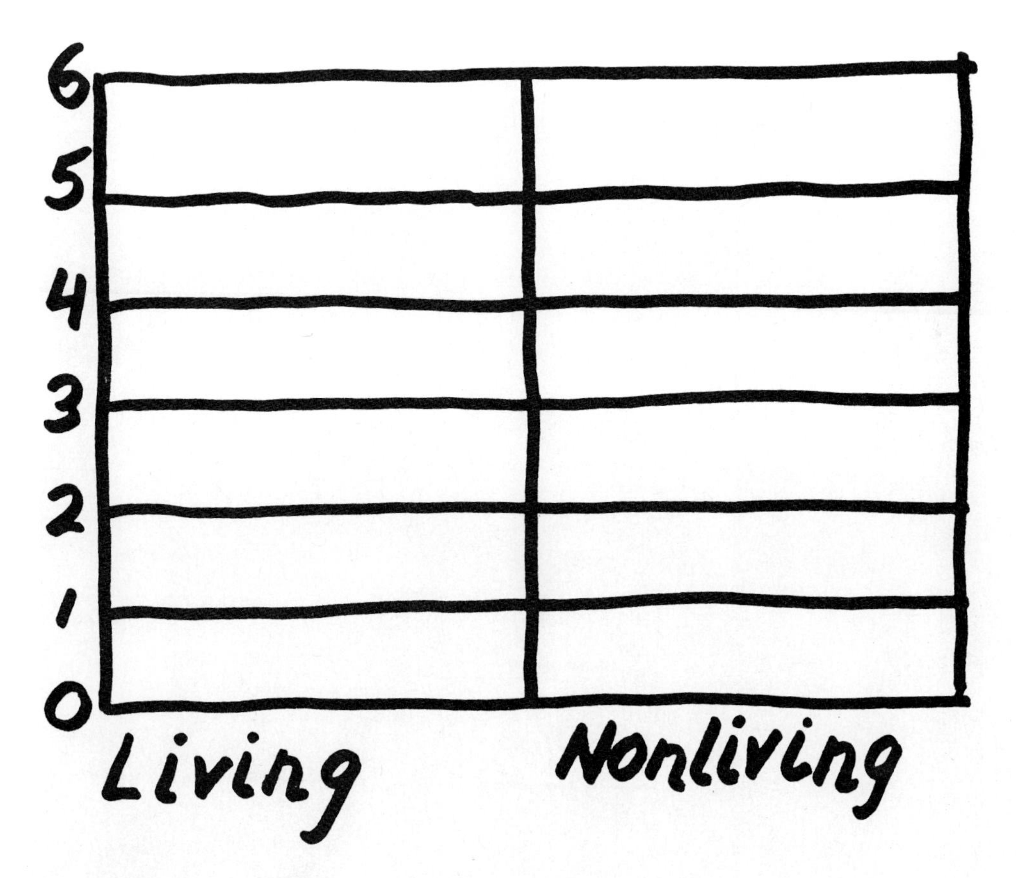

3. Observe the colors in your graph.
   What do they show about the groups?

Chapter 3 Review

Review Chapter Ideas

1. Tell what living things can do.
2. Look at the picture.
   Name the living things you see.
3. Explain what living things need.
4. Tell what people need.
5. Tell what nonliving things are like.
6. Name six nonliving things.

Review Science Words

Match the words and the pictures.

1. parents
2. shelter

a.

b.

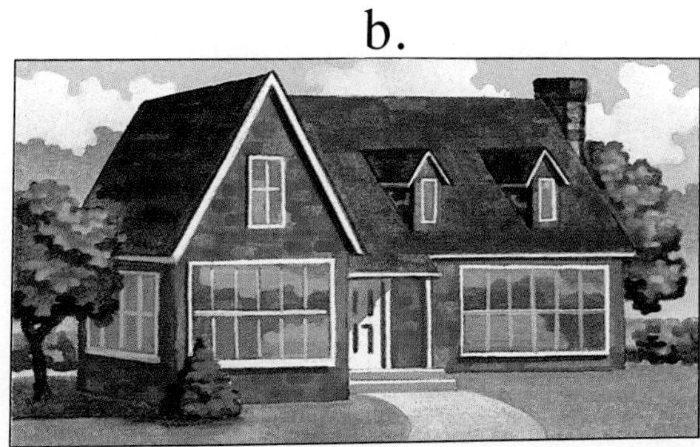

Tell what the words mean.

3. living things
4. nonliving things

Use Science Ideas

What picture does not belong?

a.

b.

c.

Chapter 4

Learning About Plants

Look at the plant in the picture.
What does the plant look like?

Starting the Chapter

Do you enjoy seeing pretty plants?

Many kinds of plants are pretty.

You can learn how these plants are alike.

Then read more about plants.

Telling About Plants

Look at these two plants.
Tell about each one.
Tell how they are alike.

Lesson 1 How Are Plants Alike and Different?

Plants are alike and different.
Plants have different colors.
Plants have different shapes.
How are these plants different?

Plants are also alike.

Most plants have the same parts.

Plants have **leaves.**

Plants have **roots.**

Find the leaves and roots here.

A plant has a **stem.**

Some plant stems are soft.

Trees are plants with hard stems.

What kinds of stems do you see here?

Lesson Review

1. What parts do plants have?
2. How are plants different?

LIFE SCIENCE

Find Out

Look outside your home.

See how plants are alike.

Grouping Leaves

Follow the Directions

1. Make two groups of leaves.
2. Find leaves with the same shape.
3. Put them in one group.
4. Find leaves with another shape.
5. Put them in a second group.

Tell What You Learned

Tell how your leaves are shaped alike.
Group your leaves a new way.

Lesson 2 How Do Plants Grow?

Many plants grow from **seeds.**
Look at the plant flowers.
Flowers of plants make seeds.
Find the seeds in this flower.

A seed opens when it begins to grow.
A plant grows from the open seed.
A growing plant makes more seeds.
Look at the seeds in the picture.
What plants will grow from the seeds?

Where can plants grow?
Plants grow in dry or wet places.
Plants grow in hot or cold places.
Where are these plants growing?

Lesson Review

1. How does a plant grow?
2. In what places can plants grow?

EARTH SCIENCE
Find Out
CONNECTION

Look at the wet place and the dry one.
How are the plants different?

Looking at Seeds

Follow the Directions

1. Find the seeds in the fruit.
2. Count the seeds in each fruit.
3. Notice how the seeds are alike.
4. Notice how the seeds are different.

Tell What You Learned

Tell how seeds are different.

What will one seed become?

Draw a picture of it.

Lesson 3 What Do Plants Need to Grow?

Plants need air to grow.

Plants also need water to grow.

Green plants need light.

How do these plants get light?

How do they get water?

Most plants grow in the ground.
They need **soil** to grow.
Soil holds plants in place.
Soil also holds water plants use.
What part of the plant grows in soil?

Plants need care to grow indoors.
People can care for indoor plants.
People give plants what they need.
What is this girl doing?

Lesson Review

1. What are four needs of plants?
2. What plants need care from people?

LIFE SCIENCE

Find Out

Find two plants that are alike.
Keep one in a dark place.
See what happens to the plants.

Bringing Water to Plants

This place is hot and dry.
Rain hardly ever falls here.
Few plants could grow here.
The plants did not get water.
Now machines pump in water.
How do the machines help plants?

What Do You Think?

How does water help dry places?

Lesson 4 Why Do People Need Plants?

People need plants for food.
People eat different plants.
Look at the farm plants here.
What parts of plants do people eat?
What plant parts do you like?

What comes from plants?

Cloth is made from parts of some plants.

Wood comes from trees.

People can make paper from wood.

Lesson Review

1. What plant parts can people eat?
2. What can people make from plants?

LIFE SCIENCE

Find Out

Look around your classroom.
Name things made from plants.

Skills for Solving Problems

Using a Hand Lens

How does a hand lens help you see?

1. Look at a leaf.

   Then use a hand lens to look at a leaf.

   Place the lens near the leaf.

   Stop when you see clearly.

   Tell what you see each time.

2. Draw two large boxes like this.

   Draw a leaf seen without a hand lens.

   Draw a leaf seen with a hand lens.

3. How does using a hand lens help you?

Chapter 4 Review

Review Chapter Ideas

1. Tell how plants are alike.
2. Tell how plants are different.
3. Explain how a plant grows.
4. Name places where plants can grow.
5. Tell what plants need to grow.
6. Look at the picture.
   Tell how the girl is using plants.

Review Science Words

Match the words and the pictures.

1. leaves
2. stem
3. roots

a. 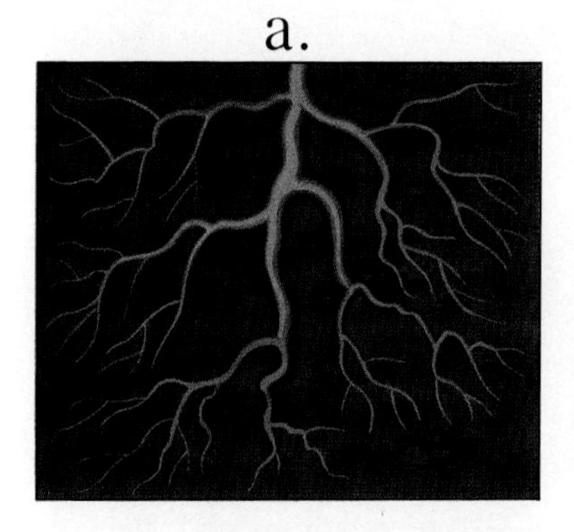

b. 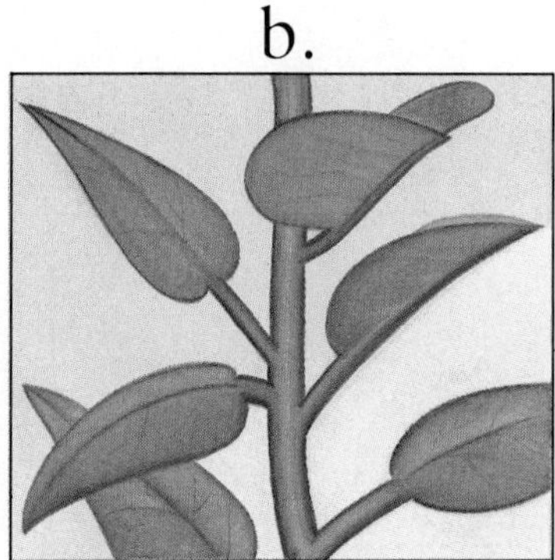

c. 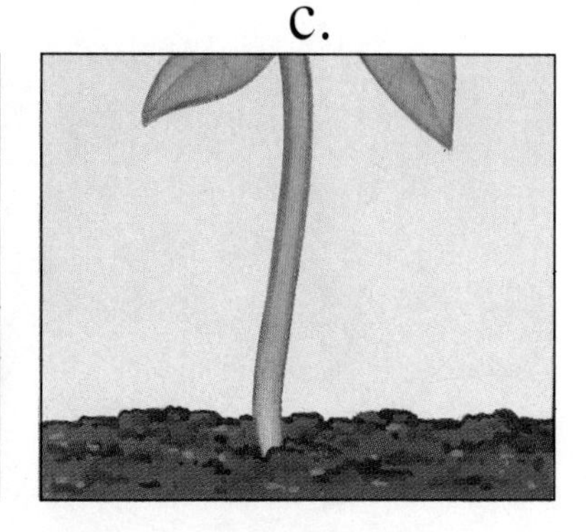

Tell what the words mean.

4. seeds
5. soil

Use Science Ideas

Tell what will happen next.

Chapter 5

Learning About Animals

Look at the animal in the picture.
What does the animal look like?

Starting the Chapter

Maybe you have touched some animals.

Point to parts of animals here.

Then read more about animals.

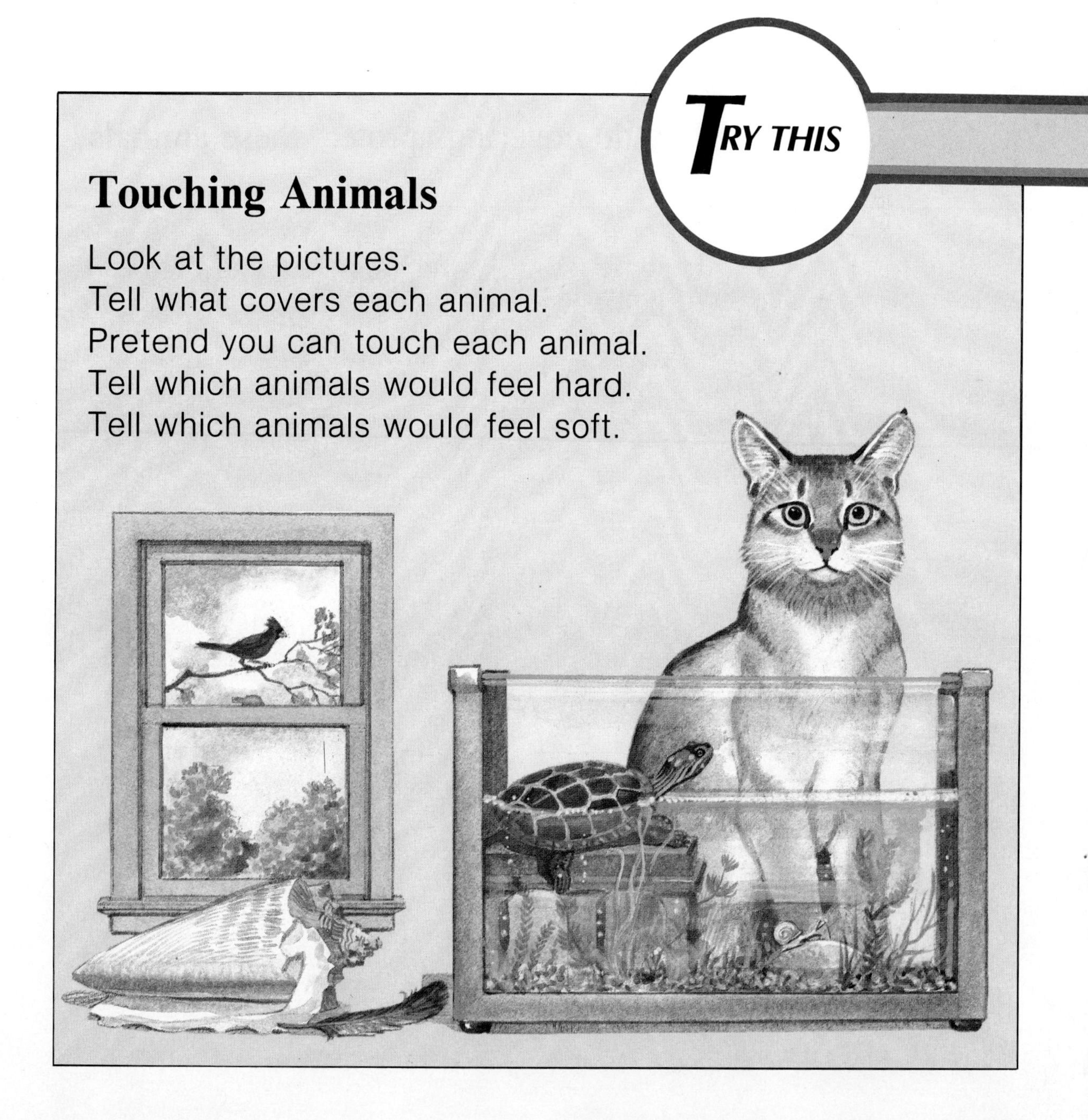

Touching Animals

Look at the pictures.
Tell what covers each animal.
Pretend you can touch each animal.
Tell which animals would feel hard.
Tell which animals would feel soft.

Lesson 1 What Ways Are Animals Different?

Animals can be big or small.
They have different shapes.
Animals have different **coverings.**
Coverings help **protect** the animals.
How do coverings protect these animals?

Animals move in different ways.
Some animals walk, fly, or swim.
Which animals walk, fly, or swim?
What ways do these animals move?

Lesson Review

1. How are animals different?
2. What ways can animals move?

PHYSICAL SCIENCE
Find Out
CONNECTION

Some coverings keep heat in bodies.
What covering keeps the bird warm?
What keeps the kangaroo warm?

Lesson 2 How Do Animals Grow?

Many animals change when they grow.
Animals change in size.
Many animals change in color.
Some animals also change in shape.
How did this butterfly change?

Baby animals have parents.
Some look like their parents.
Do these baby animals look like their parents?
Some baby animals need care to grow.
Parents protect and feed them.
What care do these baby animals get?

What baby animals do not need care?
Look at the **snake** and **insects.**
They do not need care from parents.
Snakes and insects care for themselves.

Lesson Review

1. How do animals change and grow?
2. What baby animals need care to grow?

HUMAN BODY
Find Out
CONNECTION

What did you need as a baby?
Draw a picture.
Show someone caring for you.

Observing Growing and Changing

Follow the Directions

1. Put some mealworms in a box.
2. Give them what you see here.
3. Observe the mealworms often.
4. Draw how you think they will change.
5. Observe how they grow and change.

Tell What You Learned

Draw how the mealworms changed.

Tell if they changed as you expected.

Lesson 3 Why Do People Need Animals?

People need animals for food.
People need animals for clothing.
Look at what comes from animals.
Look at the food, clothing, and shoes.
What animal does each thing come from?

People enjoy animals.

Animals can help people.

Some animals can help people work.

How does this dog help the boy?

Lesson Review

1. What things come from animals?
2. How can animals help people?

LIFE SCIENCE

Find Out

Look in magazines and books.
Find pictures of animals
helping people.

Using Something from Animals

Follow the Directions

1. Put some cold, fresh cream in a small jar.
2. Cover the jar tightly.
3. Shake it for about 10 minutes.
   Take turns shaking.
4. Observe how the cream changes.
   It is starting to become butter.

Tell What You Learned

Tell how you used cream from an animal.
Draw three ways to use butter.

Gerald Durrell

Gerald Durrell cares about animals.

He loved animals when he was young.

He began to study animals.

He collected many kinds of animals.

Later, he started a special zoo.

His zoo protects animals.

What Do You Think?

How could a zoo help animals?

Lesson 4 How Can You Care for a Pet?

People keep some animals as **pets.**
People take care of pets.
Pets need food and water.
They need a place to live.
How does this boy care for his pet?

Cats, dogs, and fish are good pets.
People enjoy having them.
What pets do you see here?
What kinds of homes do they need?

Lesson Review

1. What do pets need?
2. What animals are good pets?

LIFE SCIENCE

Find Out

Do your friends have pets?
How do they care for them?

Skills for Solving Problems

Making Charts About Animals

What can a chart show about animals?

1. Look at the pictures.

   Notice the size of each animal.

   Is the bird taller than the bear?

   Is the dog taller than the bird?

   Is the dog taller than the bear?

2. Use your own paper.
   Make a chart like this.
   Write words or draw pictures.
   Number the animals from short to tall.

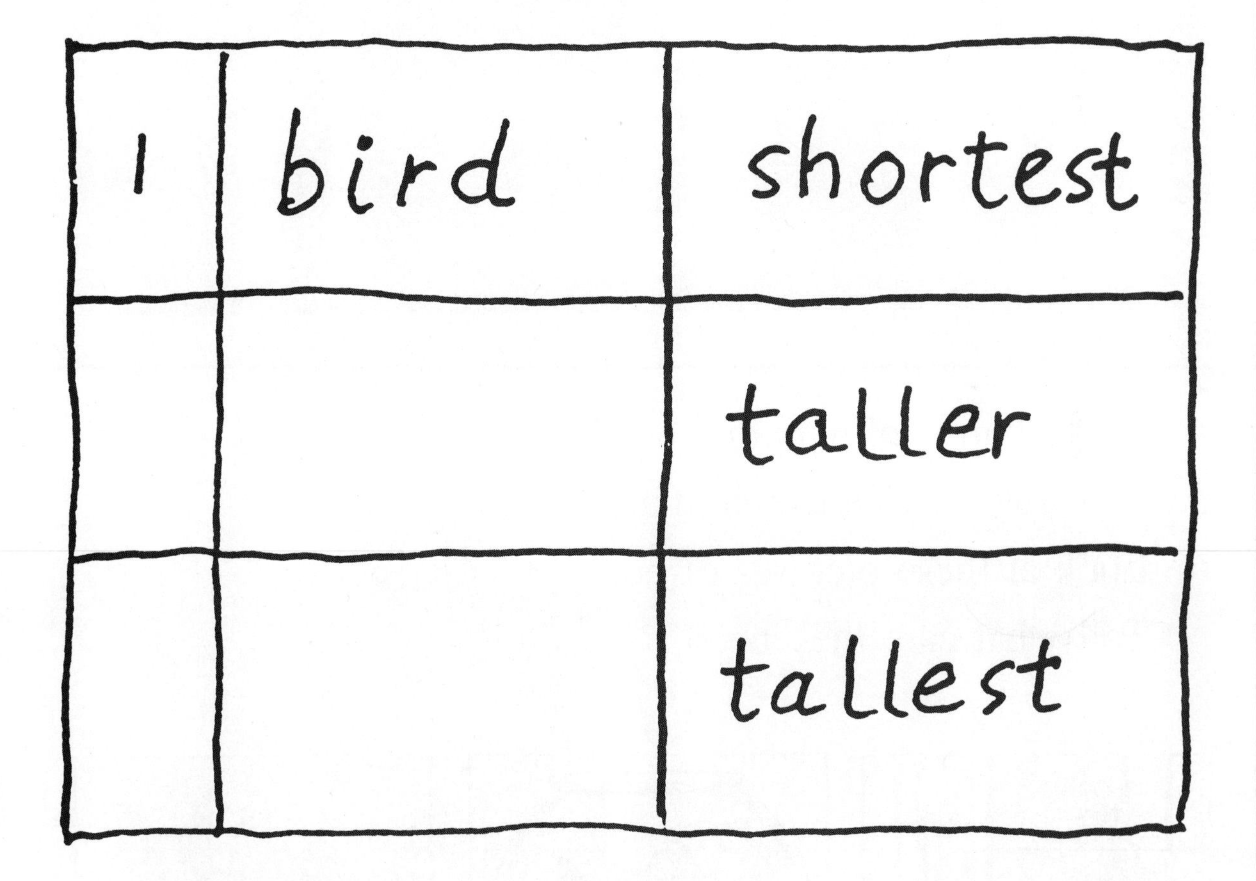

3. What does your chart show?
   Which animal is tallest?
   Which animal is shortest?

Chapter 5 Review

Review Chapter Ideas

1. Tell about differences in animals.
2. Show ways animals can move.
3. Look at these baby animals.
   Tell which look like their parents.

a.

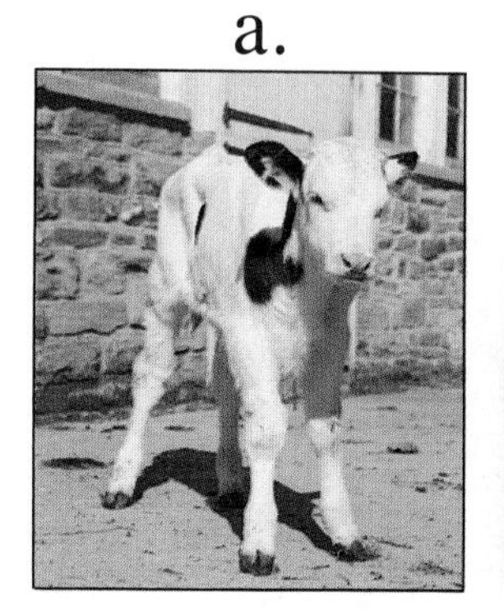

b.

c.

d.

4. Name baby animals that need care.
5. Tell how people use animals.
6. Look at these pictures of pets.
   Tell what care they need.

a.

b.

c.

Review Science Words

Match the words and the pictures.

1. insect
2. pet
3. snake

a.

b.

c.

Tell what the words mean.

4. coverings
5. protect

Use Science Ideas

Tell what happens first, next, and last.

a.

b.

c.

Pet Store Worker

Some stores sell pets.

Workers take care of the pets.

The pets stay in cages or tanks.

Workers clean the cages and tanks.

They feed the animals.

The workers know about many animals.

They tell people about the pets.

Fireflies

Why does a firefly seem to light up?

A firefly's body makes special juices.

These juices mix with air.

Then the firefly lights up.

Why does a firefly make this light?

Scientists are not sure.

It might want to attract other fireflies.

Unit 2 Review

Answer the Questions

1. Which are living things?

   Which are nonliving things?

a. b. c. d. e.

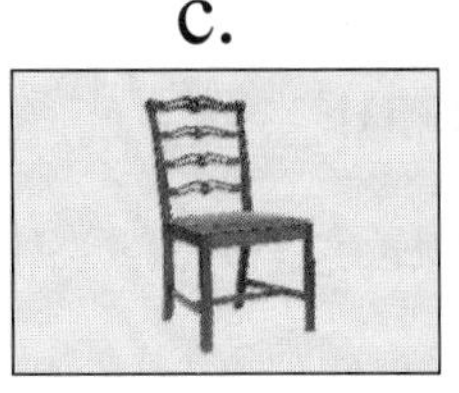

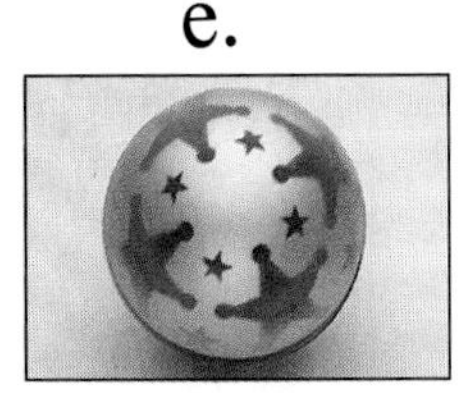

2. What parts do most plants have?
3. What is one way a plant can start?
4. What are differences in animals?
5. How do baby animals grow and change?
6. How do plants and animals help people?

Study the Picture

What do these plants need?

a.

b.

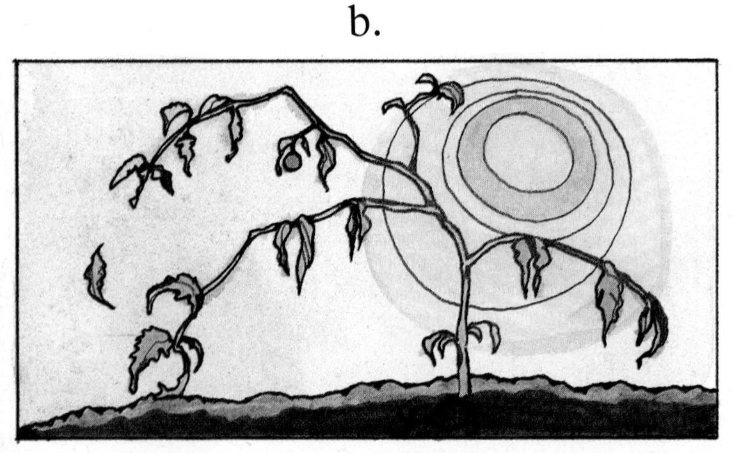

Unit 2 Projects

What to Do

1. Make a model of something alive.
   You might use clay or paper.
2. Plan a garden.
   Decide what plants you want.
   Draw a picture of your garden.
3. Make a picture zoo.
   Cut out pictures of animals.
   Paste them on a poster.

Unit 3

Physical Science

These firefighters wear special clothes.
How do they help the firefighters?
The clothes cover the firefighters.
They protect the firefighters from heat.

Pretend you are a firefighter.
What might you wear for protection?
Draw a picture of your special clothes.

Chapter 6

Grouping Things

Suppose you went to a market.
How might the food be grouped?
How would you group the food?

Starting the Chapter

What things can you put in groups?

You can group things that float.

You can group things that sink.

Then read about more ways of grouping.

TRY THIS

Grouping by Floating or Sinking

Put things like these in water.
See if they float or sink.
Make groups that float or sink.
What is in each group?

Lesson 1 What Ways Can You Group Things?

Objects are things you can see or touch.

You can group objects by shape.

You can group them by color.

How are these buttons grouped?

What other groups could you make?

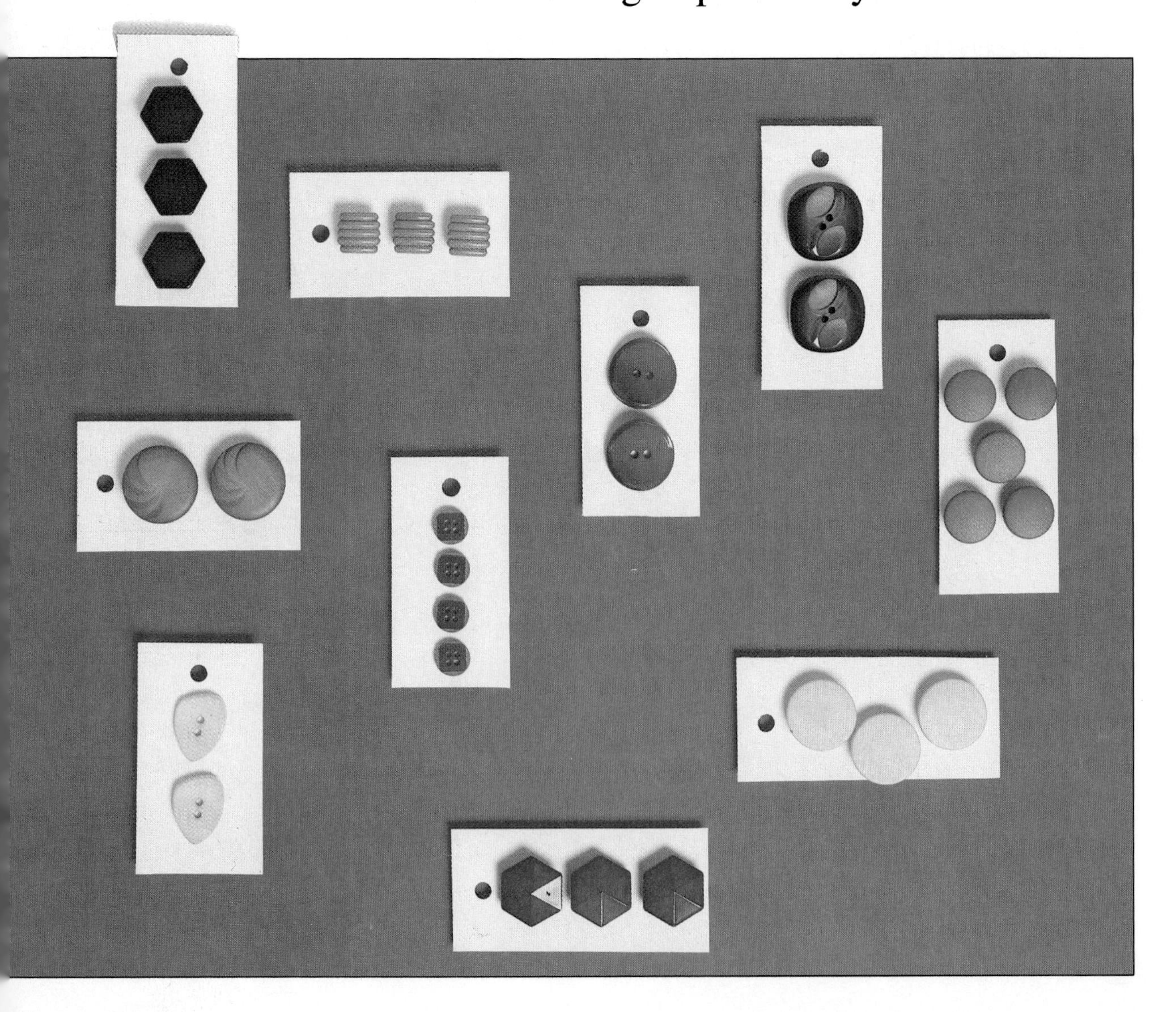

You can group objects other ways.
One way is by how long the objects are.
Another way is by how heavy they are.
How are the pens grouped?
How are the apples grouped?

3 kilograms

1 kilogram

2 kilograms

How could you group these objects?
Pretend you could touch them.
How would they feel?
What groups could you make?

Lesson Review

1. What are four ways to group things?
2. What are five ways to group rocks?

LIFE SCIENCE
Find Out
CONNECTION

How do grocery stores group fruits? Draw pictures to show it.

Grouping in Different Ways

Follow the Directions

1. Look at the objects.
2. Notice how they are alike.
3. Notice how they are different.
4. Group the objects one way.
5. Then group them another way.

Tell What You Learned

Tell one way to group the objects.
Tell another way to group them.

Lesson 2 What Takes Up Space?

Everything around you takes up **space.**
An object that is **solid** takes up space.
The book and pan are solid objects.
A **liquid** takes up space.
The juice and milk are liquids.
What other things here take up space?

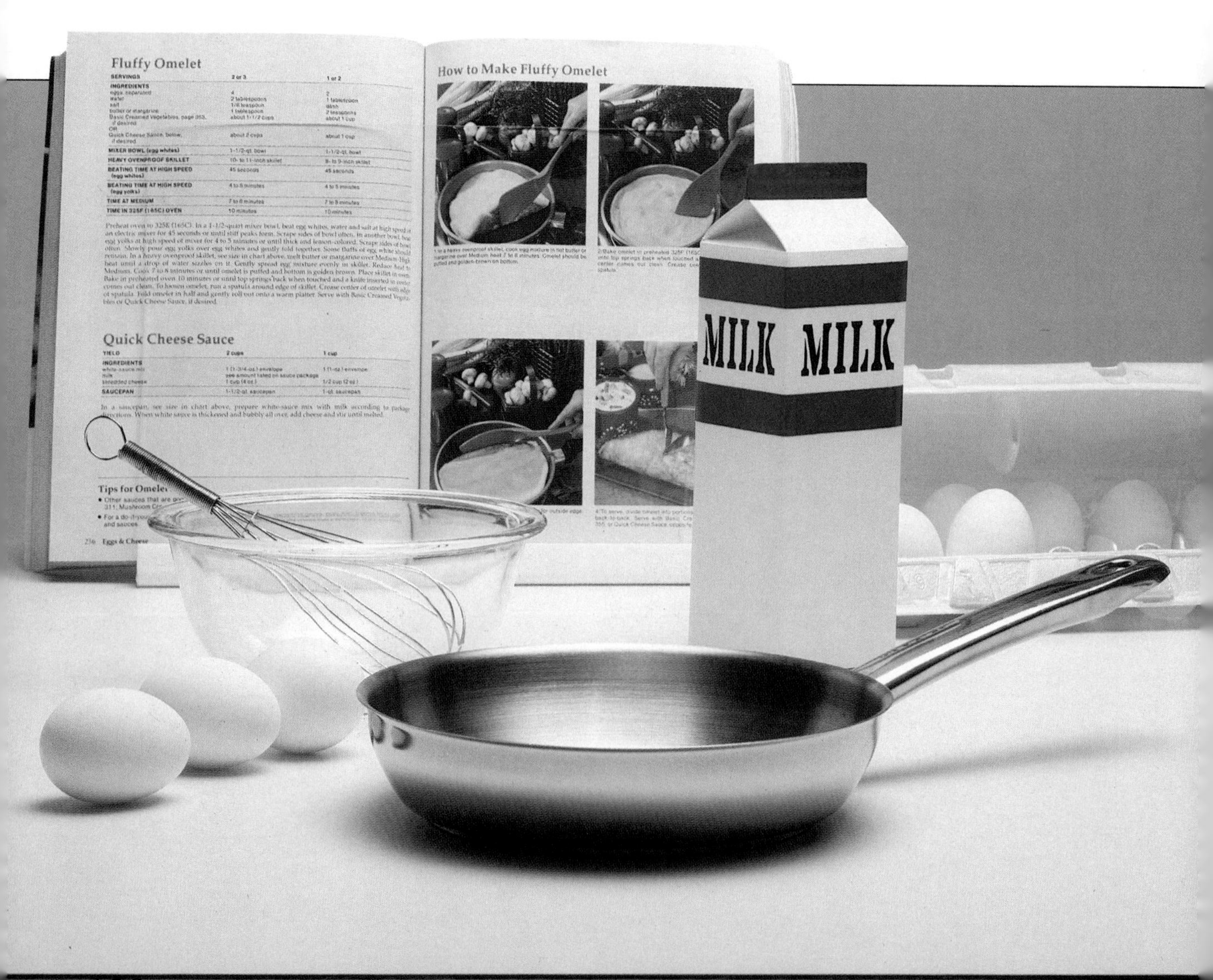

What else takes up space?

A **gas** takes up space.

Air is made of gases.

Gases take up the space in this ball.

Lesson Review

1. What things take up space?
2. Name a solid, a liquid, and a gas.

PHYSICAL SCIENCE

Find Out

What takes up the space in a shoe?

Is it a solid, liquid, or gas?

Lesson 3 What Are Solids and Liquids Like?

Solids have a certain size and shape.
Liquids have a certain size.
Liquids have no shape of their own.
They are shaped by what holds them.
What shapes do these solids have?
What shapes do these liquids have?

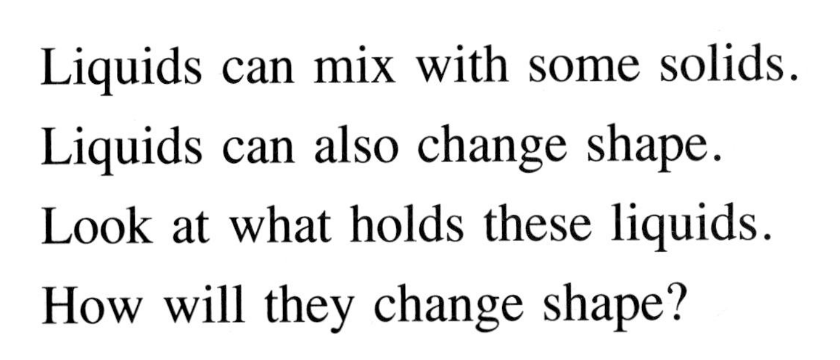

Liquids can mix with some solids.
Liquids can also change shape.
Look at what holds these liquids.
How will they change shape?

Lesson Review

1. What are solids like?
2. What are liquids like?

PHYSICAL SCIENCE

Find Out

Mix a little sugar with water.
Wait and watch.
Observe what happens.

Lesson 4 What Are Gases Like?

A gas has no special size.
A gas has no special shape.
A gas can change size and shape.
Gases filled these balloons.
How did the balloons change?

You cannot see most gases.
You can smell many gases.
You can feel a moving gas.
What would you feel here?

Lesson Review

1. What size and shape do gases have?
2. How can a gas change?

EARTH SCIENCE **Find Out** CONNECTION

Palm trees need warm weather.
What plants grow in warm places?
Look in a book about plants.

Wear cover goggles for this activity.

Blowing Up a Balloon

Follow the Directions

1. Measure the liquid.
2. Pour it into the plastic bottle.
3. Put the powder in the balloon.
4. Put the balloon on the bottle.
5. Tip the powder into the bottle.
6. Observe the balloon.

Tell What You Learned

Tell how the balloon changed.

Tell what you think filled it up.

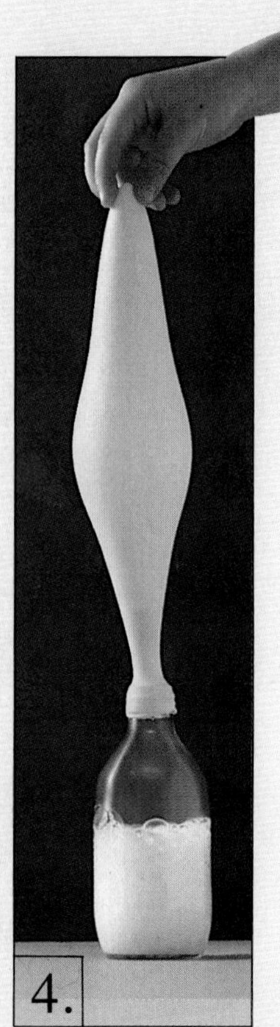

Dr. Isabella Karle

Isabella Karle built a machine.

It showed parts of things too tiny to see.

Dr. Karle used the machine to learn.

She studied tiny parts of gases.

She studied solids and liquids.

Dr. Karle made important discoveries.

What Do You Think?

How can studying help with discoveries?

Skills for Solving Problems

Measuring Solid Objects

What ways can you measure solids?

1. Measure how long each solid object is.
   Measure with a ruler.

2. Use your own paper.
   Draw each solid object.
   How long is each solid object?
   Write the numbers.

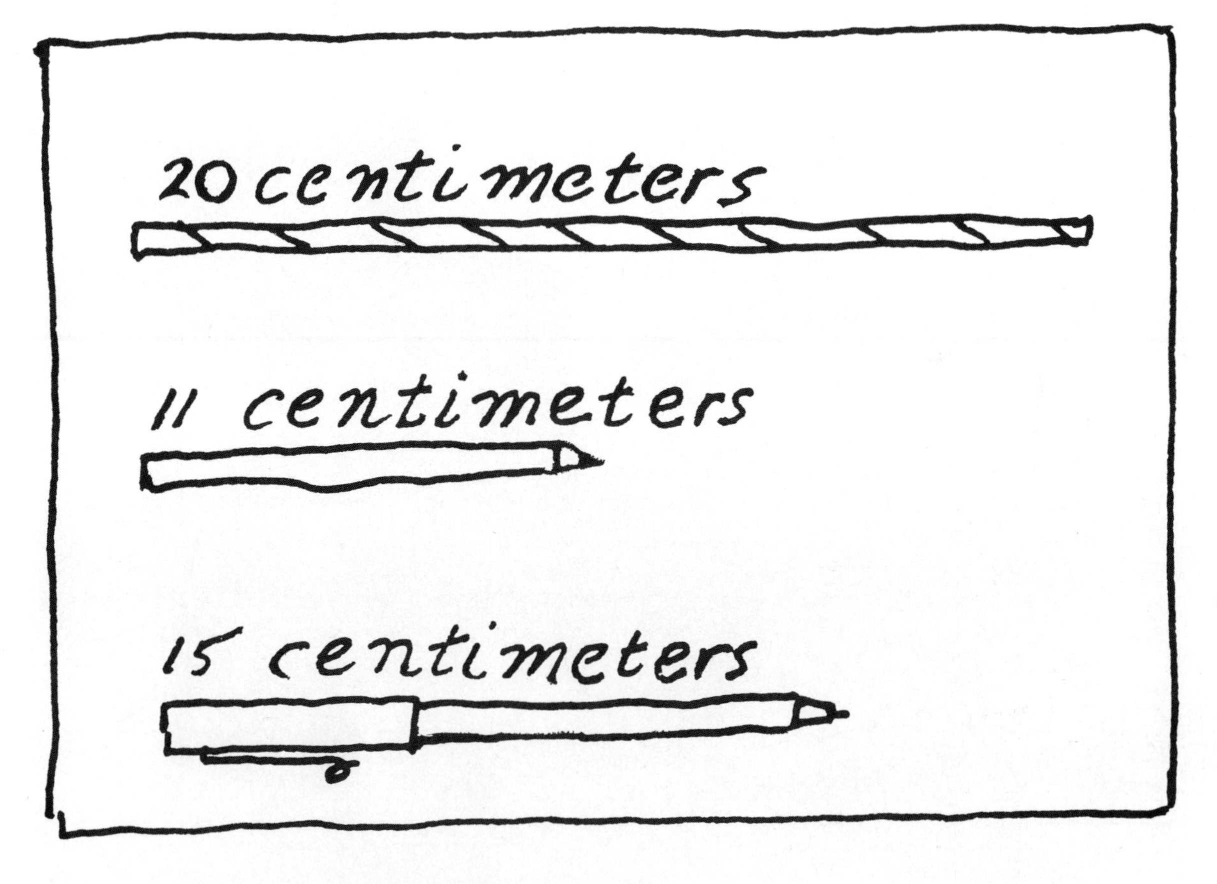

3. Look at the numbers.
   Which object is the longest?
   Which object is the shortest?

Chapter 6 Review

Review Chapter Ideas

1. Look at the picture of blocks.
   Name three ways to group them.
2. Tell what takes up space.
3. Describe a gas.
4. Describe a solid.
5. Describe a liquid.

Review Science Words

Match the words and the pictures.

1. solid
2. liquid
3. gas

a.

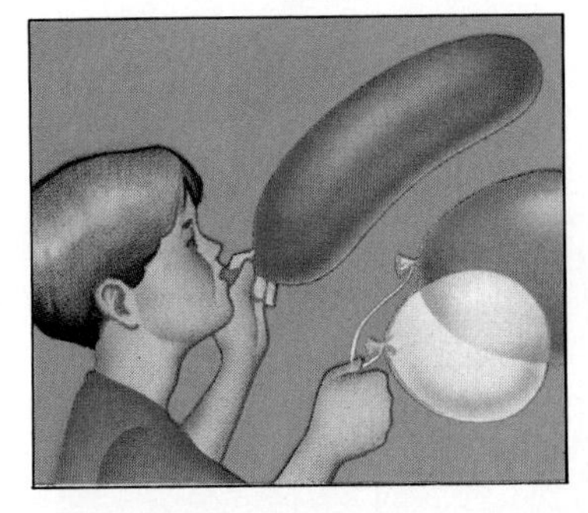

b.

c.

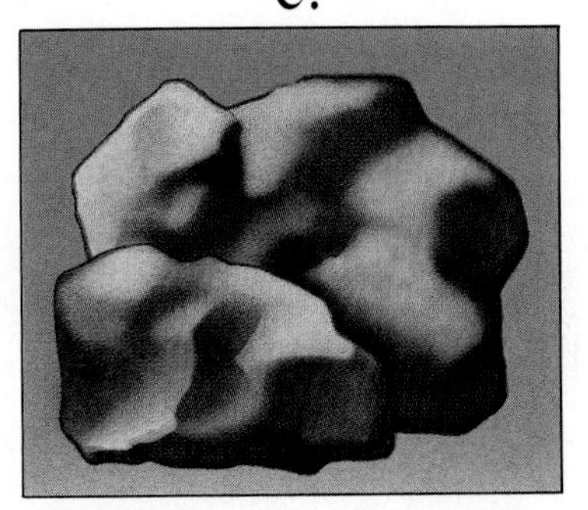

Tell what the words mean.

4. objects
5. space

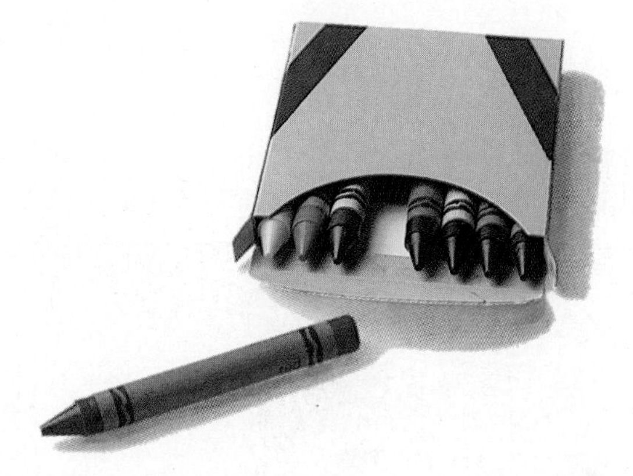

Use Science Ideas

Tell how this solid will change.

Chapter 7

Light, Sound, and Heat

Think about a warm, sunny morning.
You might see and hear many things.
What might you see and hear?

Starting the Chapter

What light do you use to read?

Try reading with more and less light.

Light helps you see better.

Then read to learn more about light.

TRY THIS

Using Different Light

Read a page using different light.
First, try reading with no light.
Next, use more light.
Last, use still more light.
Which light helps you see best?

Lesson 1 How Can Light Change?

Light comes from the sun and fires.
It also comes from an **electric light.**
Light can change.
Sometimes light is not very bright.
Light can be brighter.
How bright are these lights?

Light can change its path.
It can **bounce** off objects.
Sunlight comes in this window.
The light reaches the mirror.
Then it bounces off the mirror.
Where does the light go next?

What can block the path of light?
Objects like these block light.
Light cannot pass through them.
Objects like these make **shadows.**

1. What ways can light change?
2. What can block light?

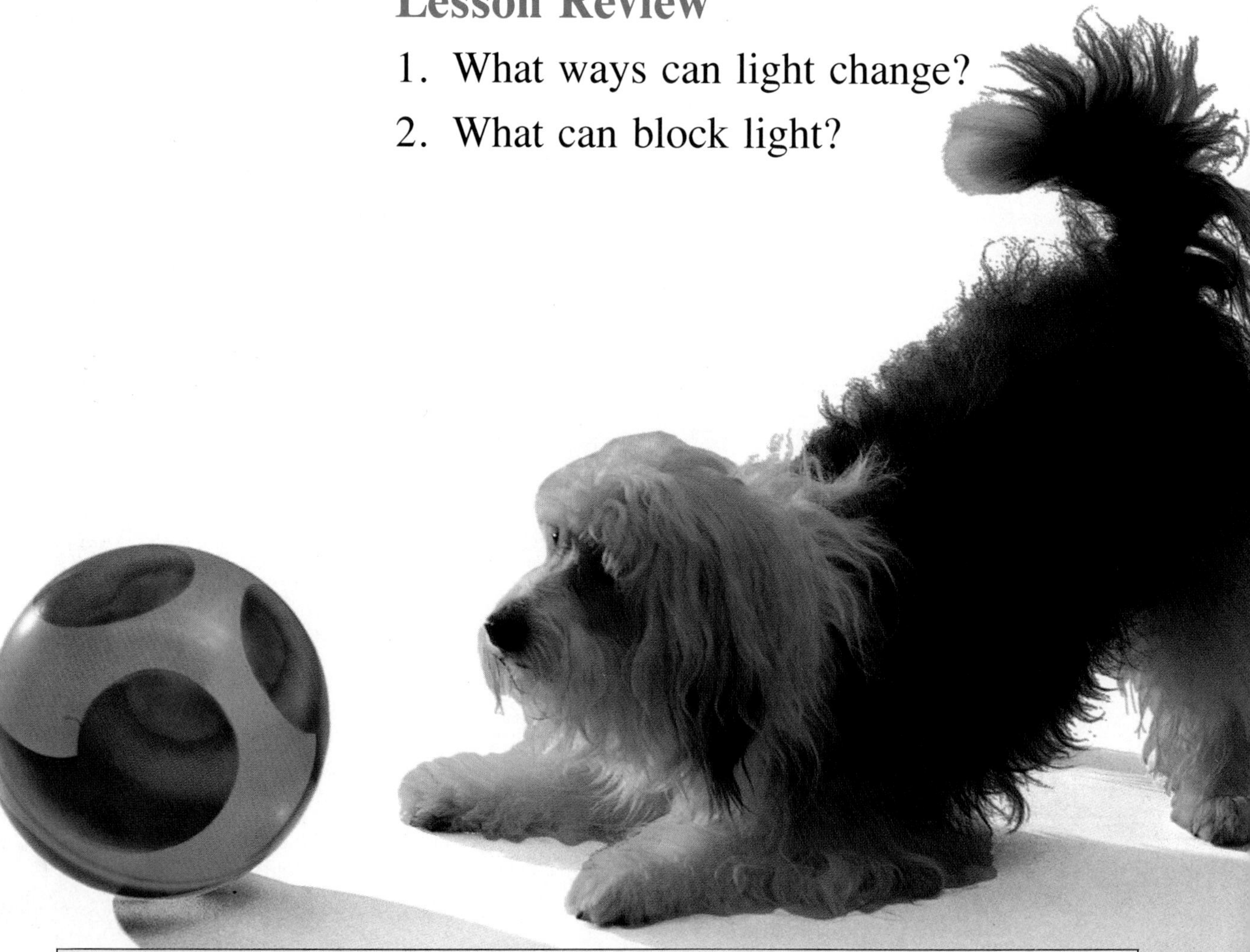

PHYSICAL SCIENCE
Find Out
CONNECTION

Look around your home.
Find things that give off light.
Draw pictures of them.

ACTIVITY

Making Shadows

Follow the Directions

1. Make shadows of objects this way.
2. Ask others to guess your objects.
3. Ask them to make shadows.
4. Guess what objects they use.

Tell What You Learned

Tell what objects made the shadows.

Tell another way to make shadows.

Lesson 2 How Can Sound Change?

Animals and people make sounds.
Many objects make sounds.
All the sounds are different.
The sounds change with each thing.
What makes sound here?

Sound can change from loud to soft.
This person plays music.
How might the sounds change?
Sound can also change **direction.**
It can go many different ways.

What can sound go through?
Sound can go through gas and liquid.
Sound also can go through solids.
What does sound go through here?

Lesson Review

1. What ways does sound change?
2. What can sound go through?

LIFE SCIENCE *Find Out* CONNECTION

Name some zoo animals.
Which animals make loud sounds?
Which make soft sounds?

Listening to Sound

Follow the Directions

1. Tie string around a spoon.
2. Hold each end of the string.
3. Put the string to your ears.
4. Tap the spoon on your desk.
5. Listen to the sound.

Tell What You Learned

Describe the sound you heard.

Tell what the sound went through.

Lesson 3 What Can You Learn About Heat?

The light from the sun carries **heat.**

Fires give off heat.

These objects give off heat.

What else makes heat?

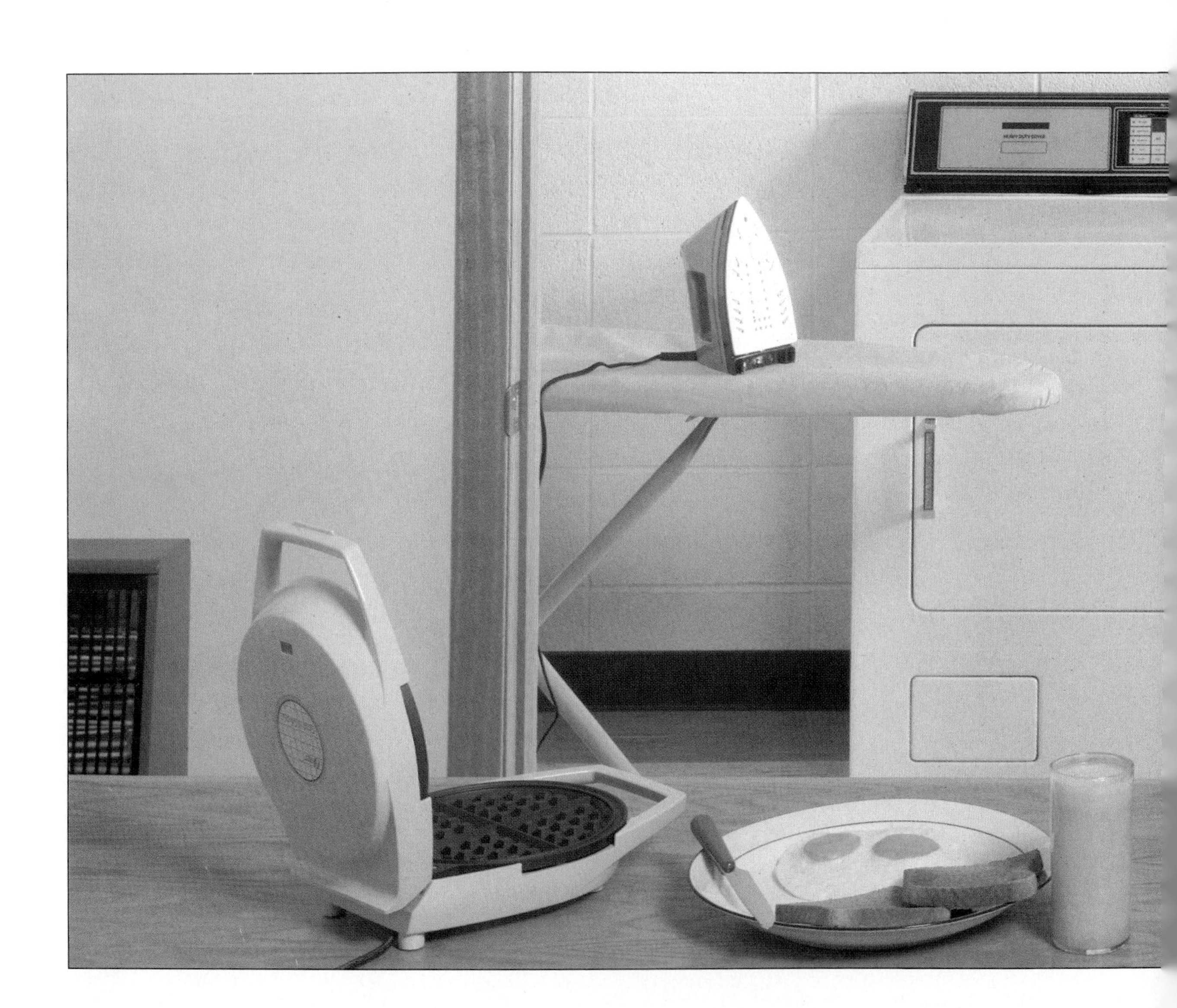

This pot is very hot.
What happens after the fire goes out?
The pot will not be as hot.
Soon the pot will only be warm.
Later, the pot will be cool.

What does a **thermometer** show?

It shows how hot something is.

What do these thermometers show?

What do other thermometers show?

Lesson Review

1. What gives off heat?
2. What shows how hot things are?

PHYSICAL SCIENCE

Find Out

What heats your home?

Draw a picture to show it.

Using a Different Thermometer

Why is this thermometer different?

You do not put it in your mouth.

You put it on your forehead.

It measures how hot your skin is.

Notice the colors on the thermometer.

Heat makes the colors change.

What Do You Think?

Why is this thermometer useful?

Using a Thermometer

What does a thermometer show?

1. Observe the thermometer.
   Notice the colors on this thermometer.
2. Put a thermometer on your desk.
   Put it in warm water.
   Put it in ice water.
   See what color it reaches each time.

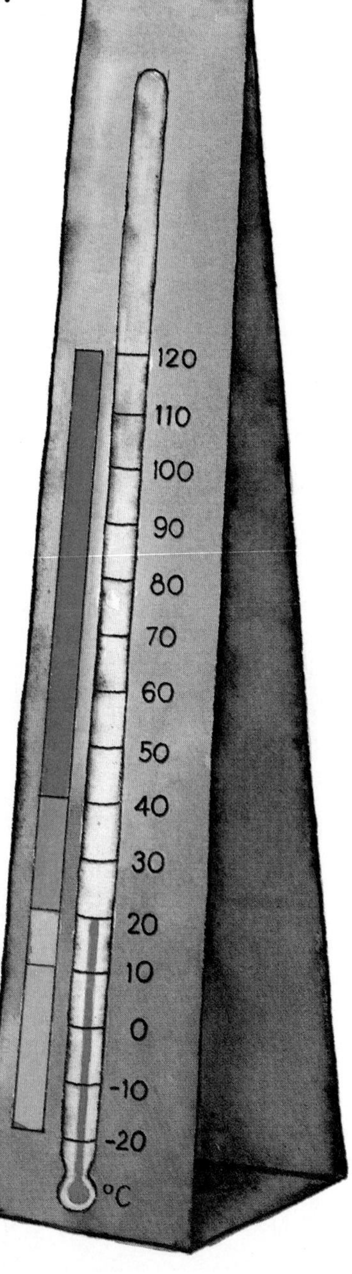

3. Use your own paper.
   Make a chart like this one.
   Show the colors.

4. Look at your chart.
   What does it show about thermometers?

Chapter 7 Review

Review Chapter Ideas

1. Tell how light changes.
2. Tell what objects make shadows.
3. Tell how sound changes.
4. Tell what sound passes through.
5. Look at the pictures.
   Tell which things give off heat.
6. Explain what a thermometer does.

a.

b.

c.

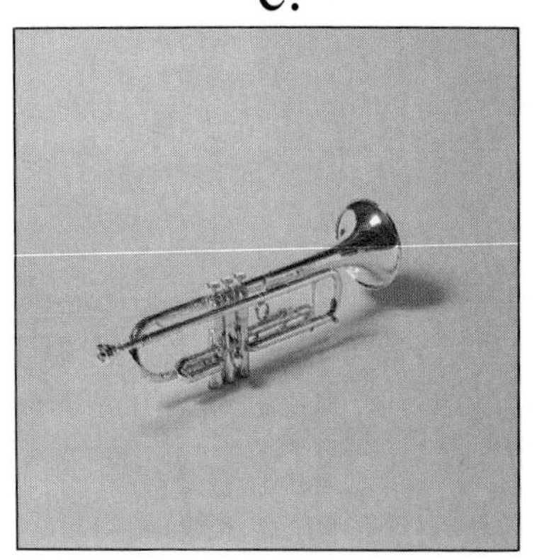

d.

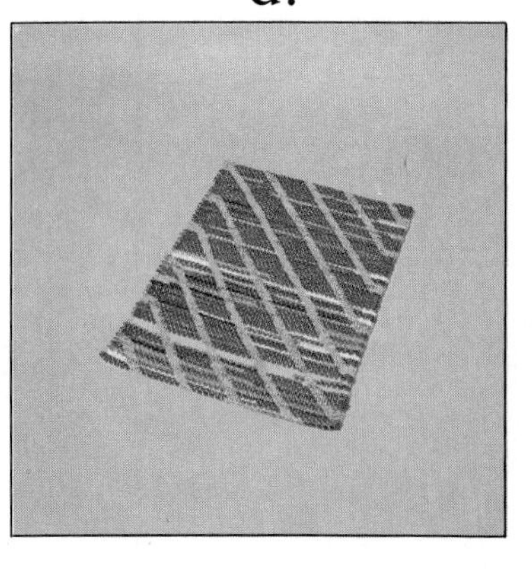

e.

f.

Review Science Words

Match the words and the pictures.

1. electric light
2. shadow
3. thermometer
4. heat

a. b. c. d.

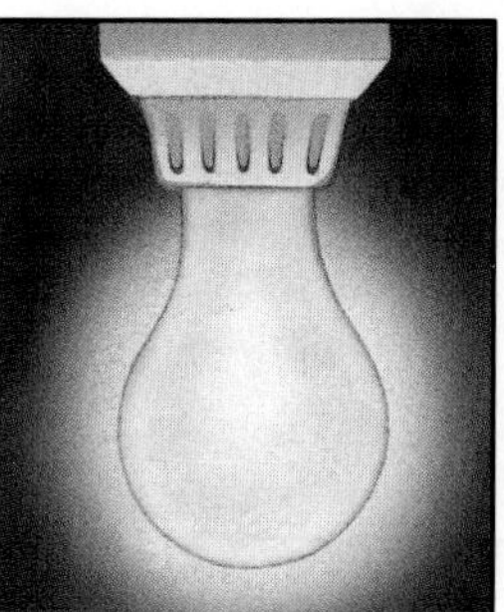

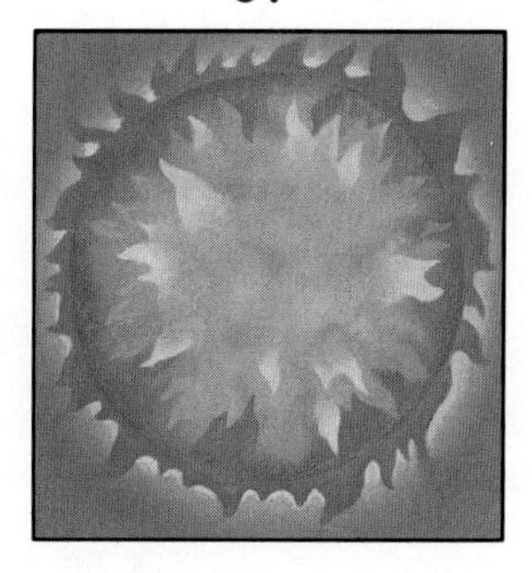

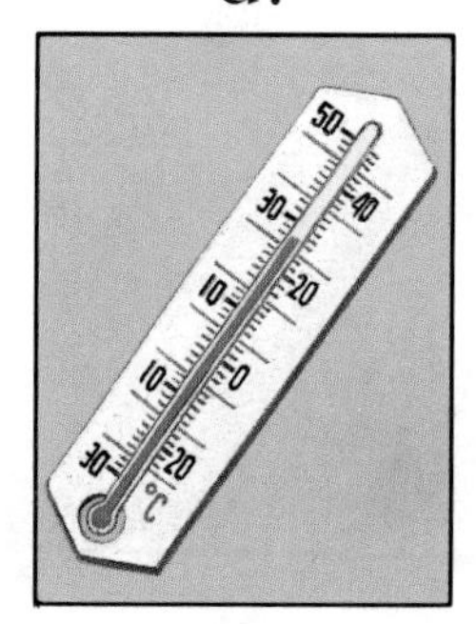

Tell what the words mean.

5. bounce
6. direction

a. b.

Use Science Ideas

Tell which car is in brighter light.

Chapter 8

Moving and Working

Airplanes move in the sky.
What work can airplanes do?

Starting the Chapter

What objects can you move at home?
What objects can you move at school?
You can make a paper airplane move.
Read more about ways things move.

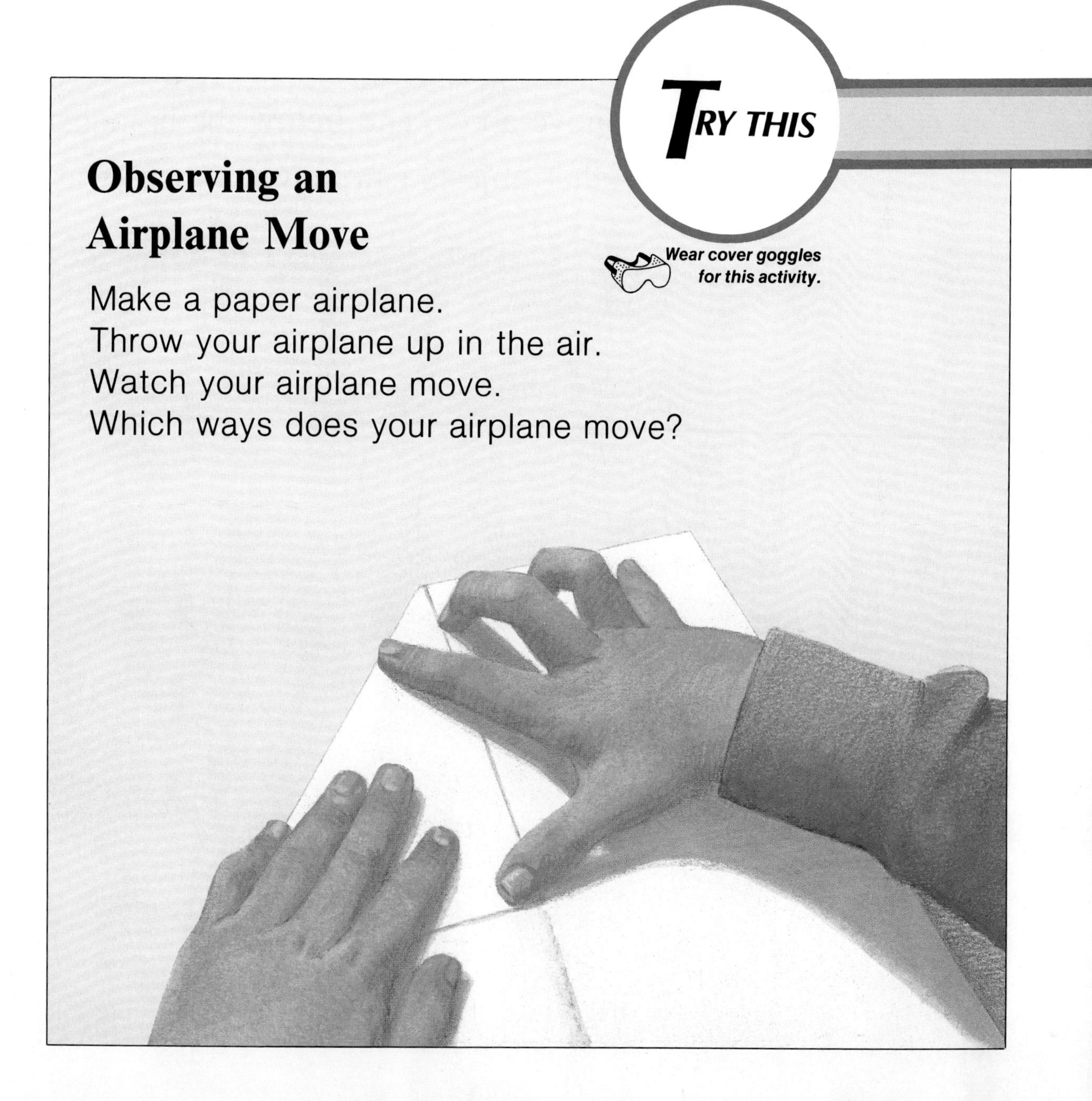

TRY THIS

Observing an Airplane Move

Wear cover goggles for this activity.

Make a paper airplane.
Throw your airplane up in the air.
Watch your airplane move.
Which ways does your airplane move?

Lesson 1 What Ways Do Objects Move?

Objects move in different ways.

The butterfly can move around the branches.

It can move up and down.

The butterfly can move left and right.

It can move above and **below** the branches.

What other objects can move in these ways?

Objects can move near or far away.
Objects and people can change **distance.**
These people move near the elephants.
Next, the people will move away.
They will be far from the elephants.
Then they will move near the birds.

Objects move at different **speeds.**
They can move fast or slow.
Which child moves fastest?

Lesson Review

1. What ways can objects move?
2. At what speeds do objects move?

HUMAN BODY
Find Out
CONNECTION

Watch friends on the playground.
Notice the ways people move.
Draw stick pictures to show it.

Using Robots for Work

People build robots that can move.

Robots move in different directions.

They can do work for people.

Robots can bring food to people.

They can help put cars together.

Someday a robot might do work for you!

What Do You Think?

What could a robot do best?

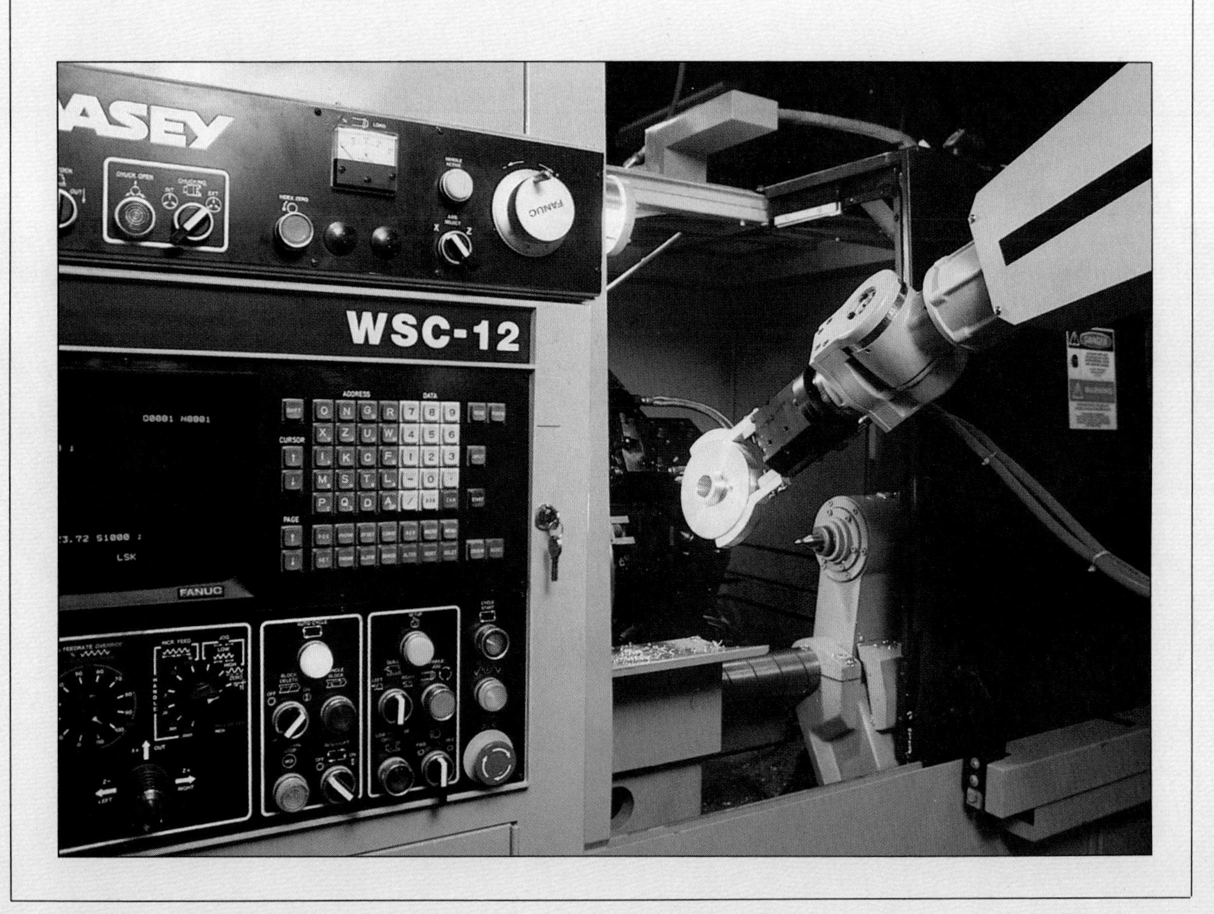

Lesson 2 What Can Move Objects?

Pushing moves objects.
Pulling moves objects.
Pushing and pulling can move a wagon.
Heavy objects are harder to move.
You must push or pull harder.
Which wagon is harder to move?

Lifting can move objects.
Objects also move when they fall.
Objects fall down toward the ground.
What is this child moving by lifting?
What direction do the objects move?
What direction will they go if they fall?

Magnets can move some objects.
Magnets attract some objects.
What objects does this magnet attract?
What shape is this magnet?

Lesson Review

1. What can make an object move?
2. How do objects move when they fall?

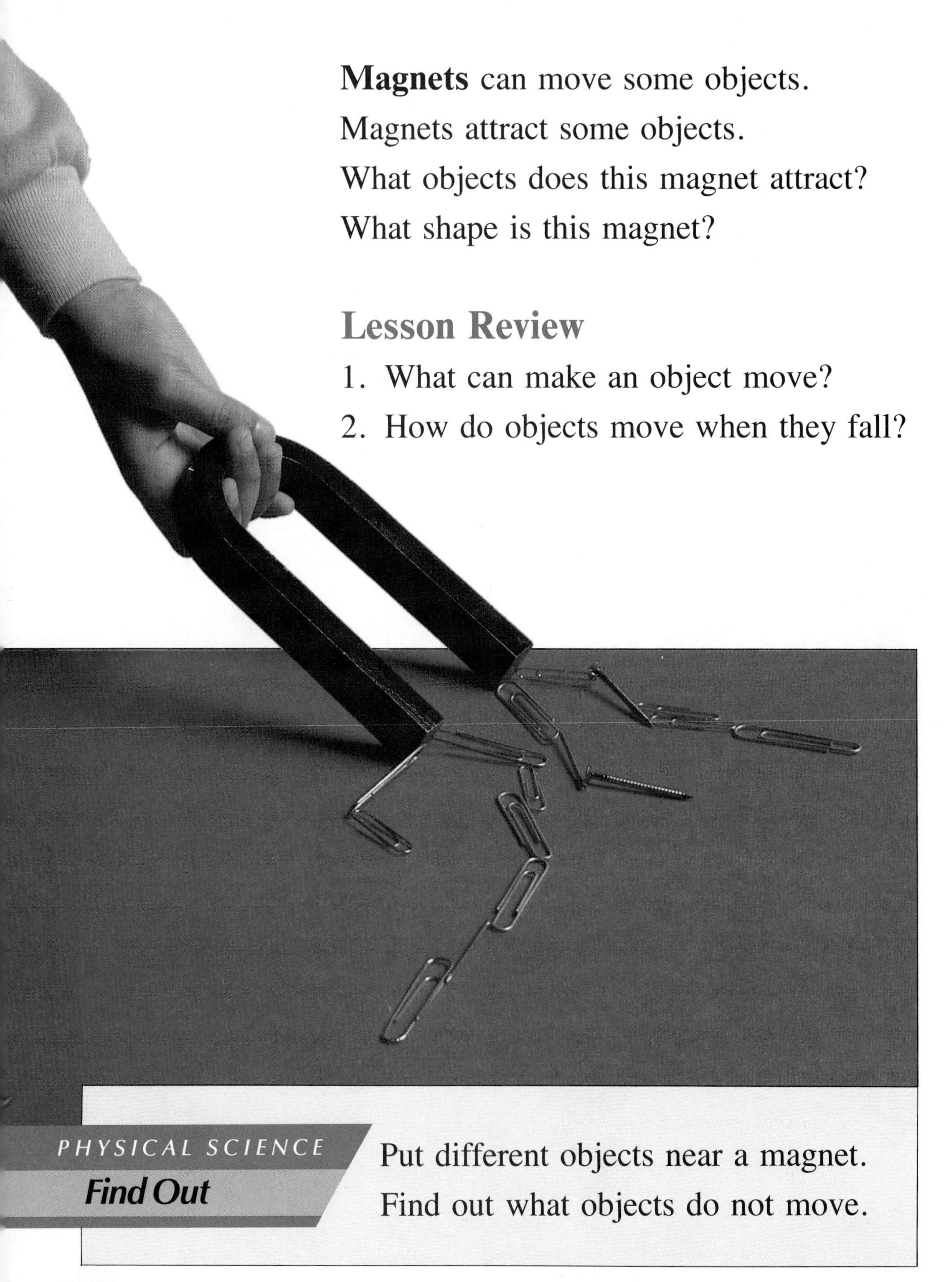

PHYSICAL SCIENCE

Find Out

Put different objects near a magnet.
Find out what objects do not move.

Using a Magnet

Follow the Directions

1. Cut out a small paper circle.
2. Tape a paper clip to the circle.
3. Put the circle on a paper plate.
4. Keep the paper clip next to the plate.
5. Move a magnet under the plate.
6. Observe what the circle does.

Tell What You Learned

Tell what you used a magnet to do.

Tell another way to use a magnet.

Lesson 3 What Work Can Machines Do?

What is work?

Work is using a push to move objects.

It is using a pull to move objects, too.

Machines that move objects do work.

Which machine here pushes?

Look at these different machines.
They help people.
The machines are used to lift and carry.
What work does each machine do?
When do you use machines?

People use machines for farming.
They use machines for building.
People use machines in homes.
How do people use machines here?

Lesson Review

1. What kinds of work can machines do?
2. How do machines help people?

LIFE SCIENCE
Find Out
CONNECTION

Look around your home. What machines can help prepare foods?

Moving a Rock

Follow the Directions

1. Put together a machine like this.
2. Work with a partner.
3. Move a rock using the machine.
4. Notice what direction you pull the string.
5. Notice what direction the rock moves.

Tell What You Learned

Tell what work the machine did.

Tell another way to use this machine.

Skills for Solving Problems

Reading a Timer

What machine makes work easiest?

1. Look at the three pencil sharpeners.
   Find the timers above each one.
   The first machine sharpened 6 pencils.
   It took 1 minute and 13 seconds.
   How many pencils did the others sharpen?
   How long did it take each machine?

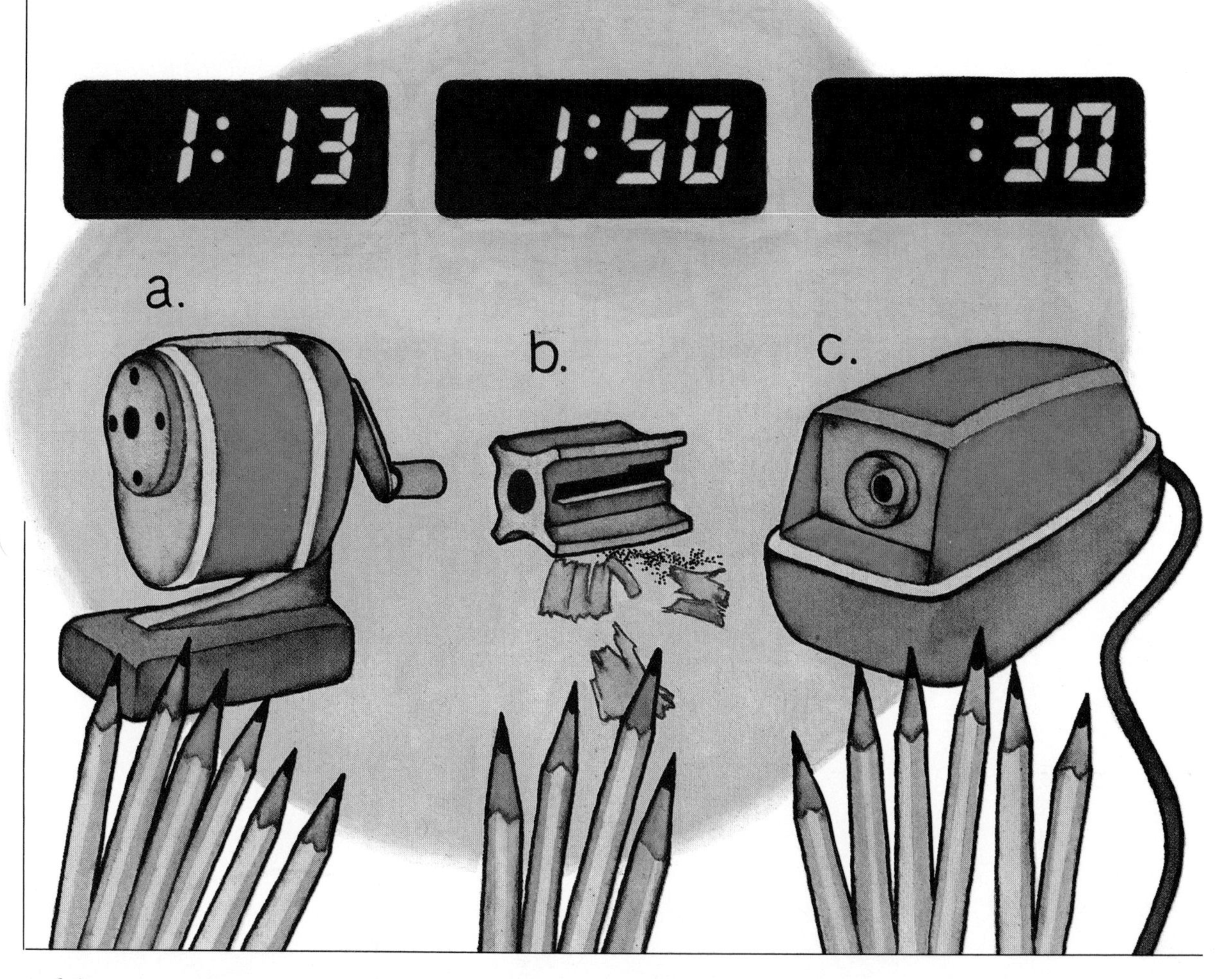

2. Make a chart like this one.
   Write the number of pencils sharpened.
   Tell if it took more than 1 minute.

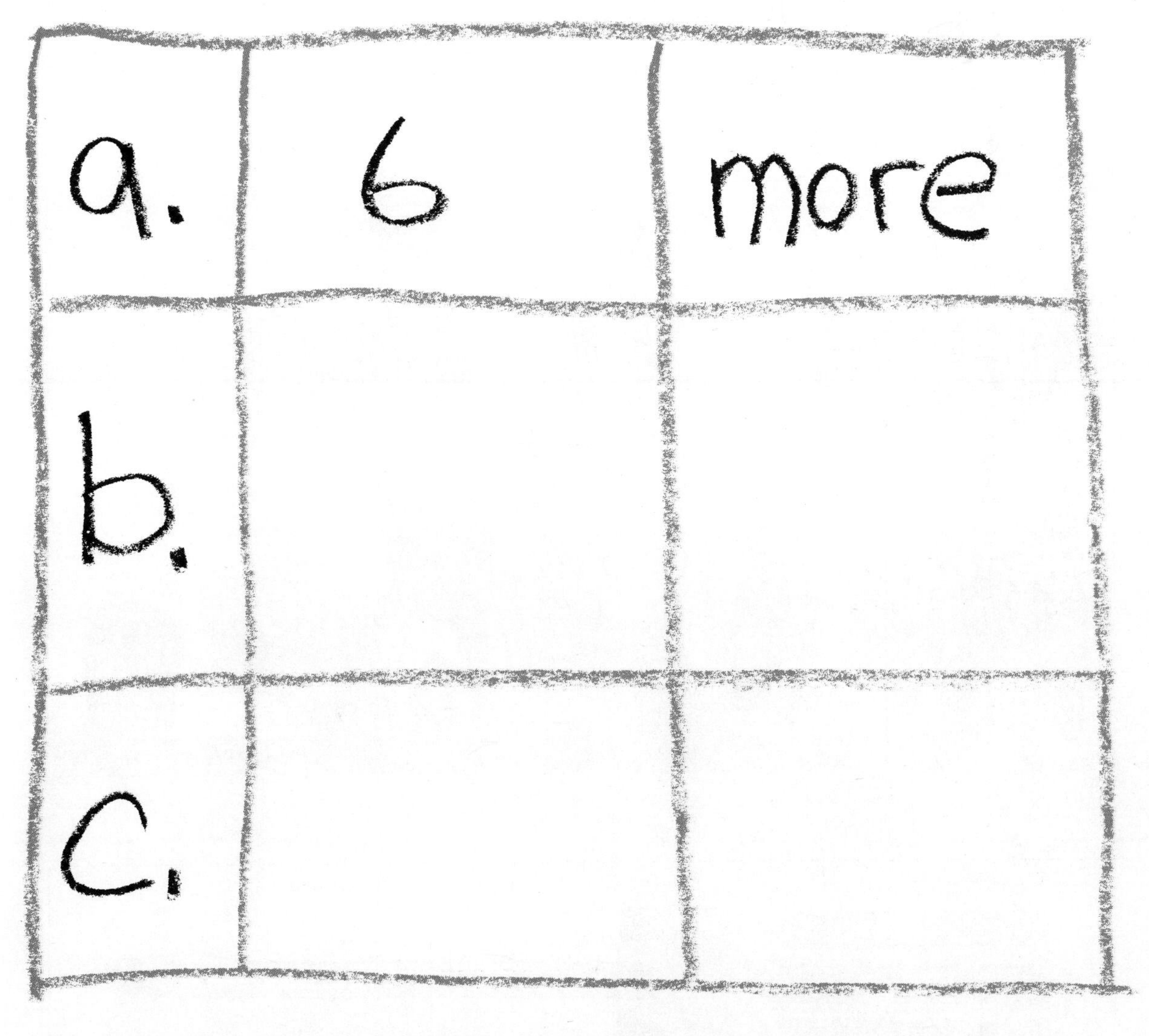

3. Which machine works the fastest?

Chapter 8 Review

Review Chapter Ideas

1. Tell some ways things can move.
2. Tell how objects change speed.
3. Tell what can move objects.
4. Describe what a magnet can do.
5. Look at the pictures.

   Tell how machines help people.

a.

b.

c.

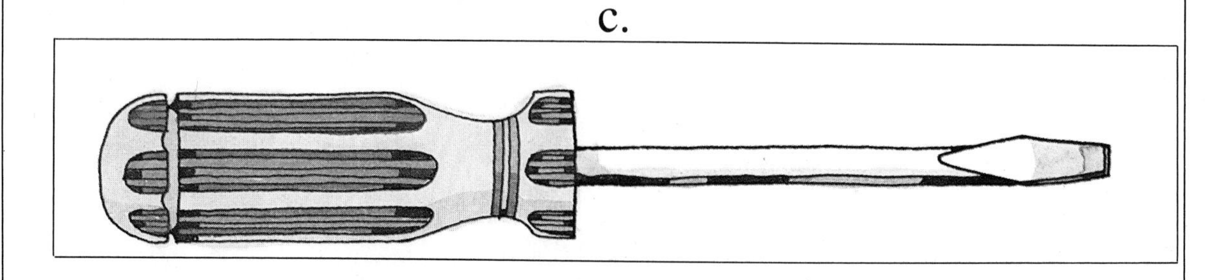

Review Science Words

Match the words and the pictures.

1. magnet
2. machine
3. below

a.

b.

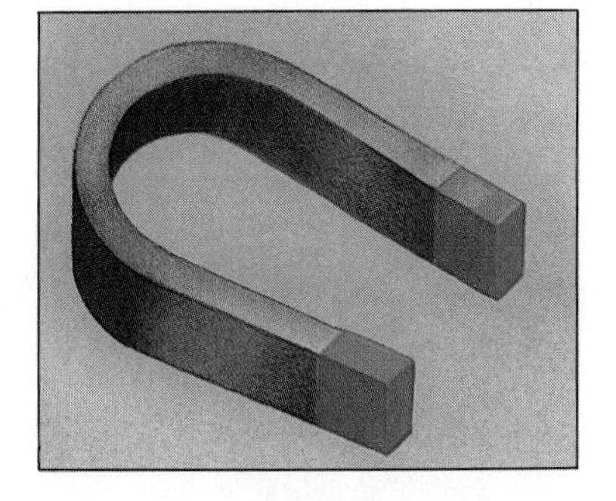

c.

Tell what each word means.

4. distance
5. speed

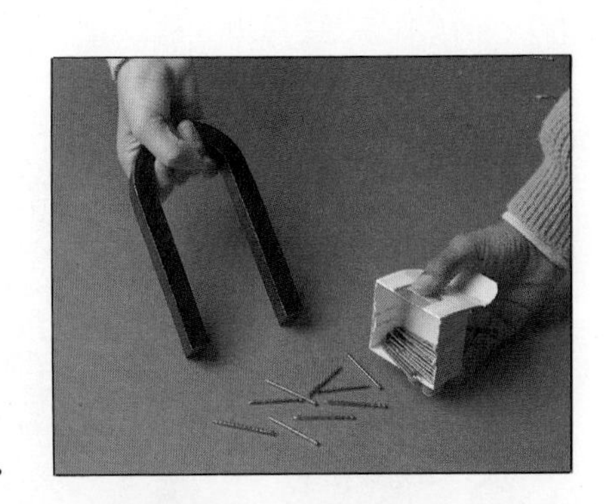

Use Science Ideas

Tell what you think will happen next.

Pilot

Airplanes are big machines.

They are used to move people and objects.

Pilots fly many kinds of airplanes.

Pilots must know how air moves.

They check airplanes for safety.

Pilots make airplanes move many ways.

Pencil Sharpener

A pencil sharpener has two rollers.
The rollers have sharp edges.
A pencil fits between the rollers.
The space is wide at one end.
It comes to a point at the other end.
The rollers spin when you turn the crank.
The rollers shape your pencil.
Then your pencil has a point.

Unit 3 Review

Answer the Questions

1. What is each thing like?

a. 

b. 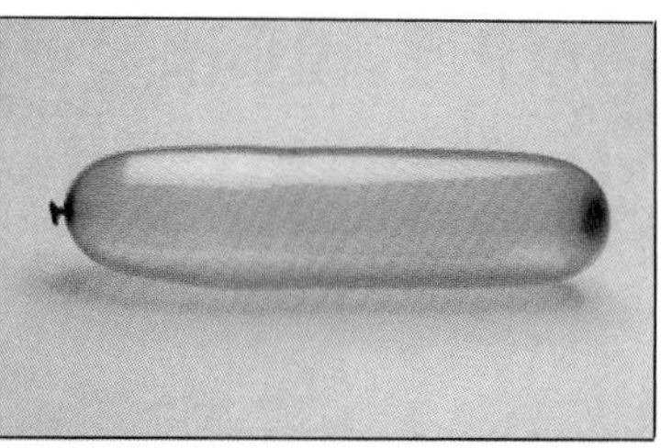

c. 

2. What ways can light change?
3. Where does heat come from?
4. What ways can sound change?
5. What ways can objects move?
6. What makes objects move?

Study the Pictures

Tell what machine does not belong.

Why does it not belong?

a. 

b. 

c. 

What to Do

1. Make liquid for blowing bubbles.

   Use water and liquid soap.

   Blow bubbles.

   Try to keep them in the air.

2. Keep your fingers as you see here.

   Make different sounds.

   Hum. Sing.

   Find out what you can feel.

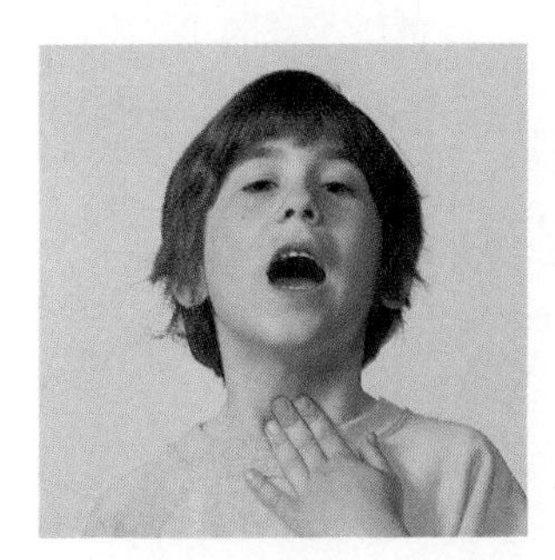

3. Pretend you have no machines.

   Tell a story.

   Tell about a day with no machines.

Unit 4

Earth Science

Pretend you live on the moon.
You could see the earth.
It might look like this.

You live on the earth.
You can see land, water, and sky.
How does the sky look today?

Chapter 9

The Earth

The earth has many beautiful places.
What places does the picture show?

Starting the Chapter

The earth has land and water.

You live on the land.

See how much water the earth has.

Then read to learn more about the earth.

TRY THIS

Observing Land and Water

Look at a globe of the earth.
The blue part is water.
The other parts are land.
Does the earth have more land or water?

Lesson 1 What Does the Earth Have?

The earth is round like a ball.
The earth has air all around it.
The earth has water and land.
The land has flat places.
It has **mountains** and **valleys.**
How does the land look here?

Land is made of many kinds of **soil.**
Land is made of many kinds of rock.
Most mountains are made of rock.
What colors of rock do you see here?

What can people get from the land?
They can get **coal** and **oil.**
People can try to use less coal and oil.
Then coal and oil can last longer.

Lesson Review

1. What three things does earth have?
2. What things can people get from land?

LIFE SCIENCE
Find Out
CONNECTION

Look at some soil.
Do you see any living things?
What nonliving things do you see?

Grouping Rocks

Follow the Directions

1. Look at different kinds of rocks.
2. Notice the colors, sizes, and shapes.
3. Touch each rock.
4. Then group the rocks.
5. Group them by how they are alike.

Tell What You Learned

Tell how the rocks are alike.

Think of other ways to group the rocks.

Lesson 2 Where Is the Water on Earth?

Most water on Earth is in **oceans.**
Water is in lakes and rivers, too.
It is also in **streams** like this one.
The earth gets water when it rains.
What comes from melting ice and snow?

The ocean has salty water.
People cannot drink this salty water.
They can drink only fresh water.
Where do these people get fresh water?

Lesson Review

1. Where do you find water on earth?
2. What kind of water do people drink?

EARTH SCIENCE

Find Out

What water is near you?
Is it salty water or fresh water?

Lesson 3 How Is Air Useful?

People need air to breathe.
You cannot see clean air.
You usually cannot smell or taste air.
Where is air in this picture?

Moving air is wind.
What can wind do?
You need wind to fly a kite.
Warm, moving wind can dry clothes.
Wind moves the top of this machine.
Wind helps the machine pump water.

Smoky fires put dirt into the air.
These cars put dirt into the air, too.
Using cars less helps keep the air clean.
How else can people keep the air clean?

Lesson Review

1. How do people use air?
2. What helps keep air clean?

PHYSICAL SCIENCE
Find Out
CONNECTION

Think about some machines.
What machines help move air?

Using Air to Move Boats

Follow the Directions

1. Make a boat like the one here.
2. Put your boat in water.
3. Blow air on your boat.
4. Make it move on the water.

Tell What You Learned

Tell how the boat moved.

What other ways can we use air?

Lesson 4 How Do People Use Land and Water?

People use plants they grow on the land.
They use coal from deep in the land.
People use oil they pump from the land.
They build homes and farms on land.
How do people use this land?

People drink water.
They use it for washing.
People swim and fish in water.
They also travel on water by boat.
What is water used for here?

These people are careful.
They keep the land and water clean.
They use the land and water wisely.
What can you do to help?

Lesson Review

1. How can people use land and water?
2. What is careful use of land and water?

EARTH SCIENCE

Find Out

Trees grow on land.
How can people use trees wisely?

Rachel Carson

Rachel Carson cared about the earth.
She studied and learned about oceans.
She wrote books about them.
Some books told how she loved oceans.
Her books also helped many people.
They learned to care about oceans, too.

What Do You Think?

How can people care about oceans?

Using a Map

What can a map show?

1. Look at this map.
   Find the land.
   Point to the oceans.
   Find a river.

2. Choose a place you have seen.
   Use your own paper.
   Draw a map like these maps.
   Write what each thing on your map is.

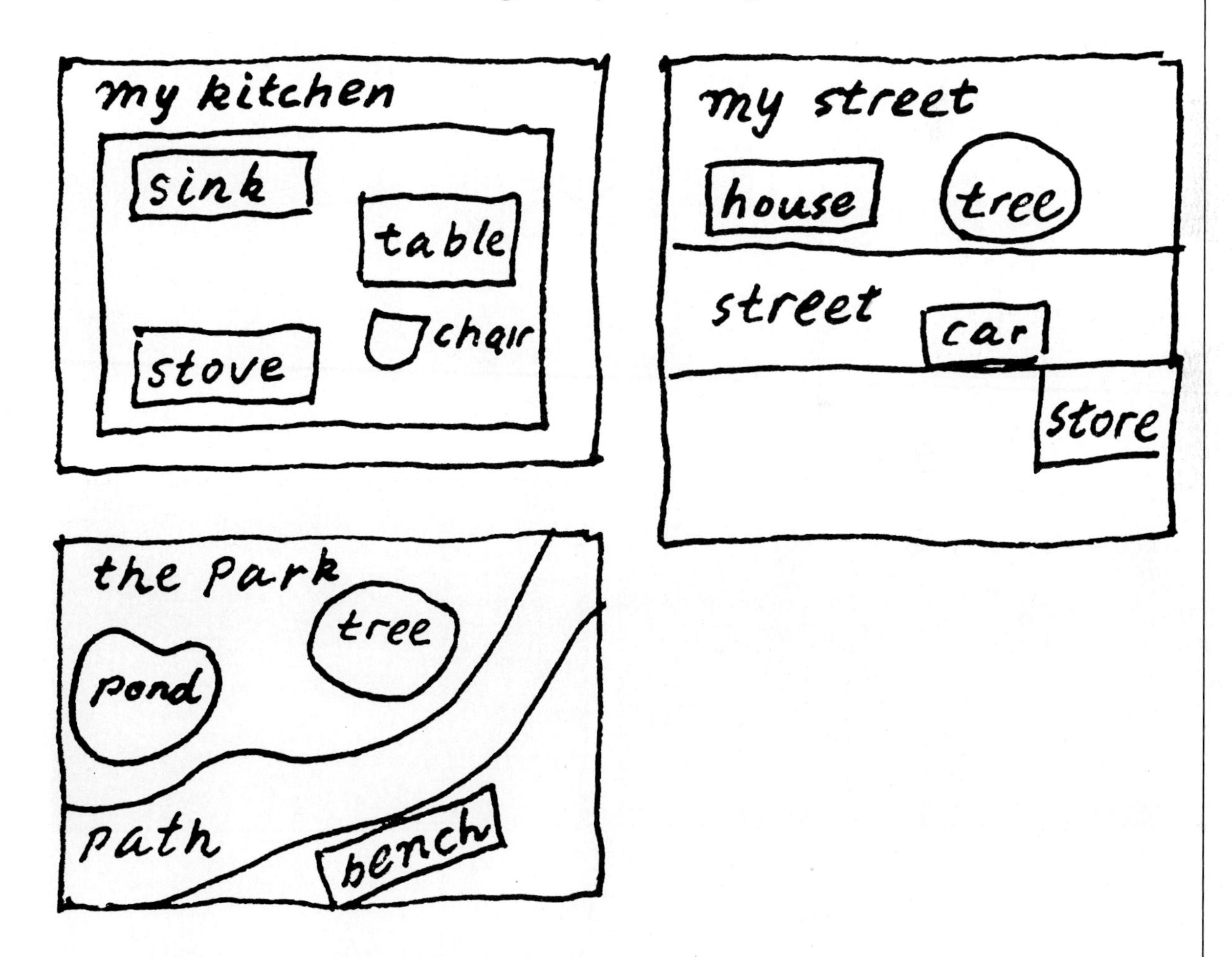

3. Show your map to a friend.
   Point to places on your map.

Chapter 9 Review

Review Chapter Ideas

1. Tell what the earth has.
2. Describe what the land has.
3. Look at the pictures.
   Tell about the water on earth.

a.

b.

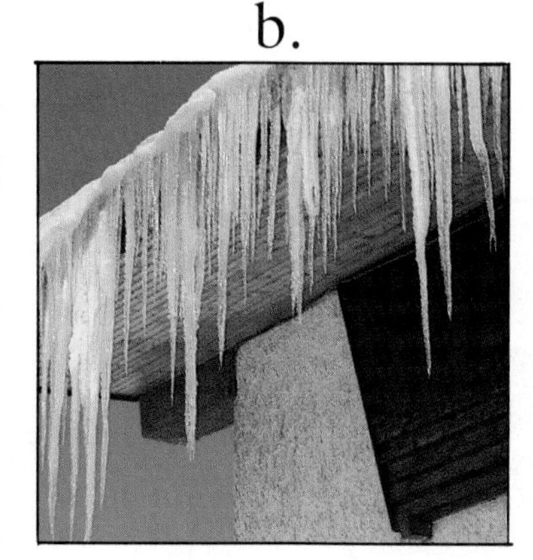

c.

d.

e.

4. Describe what air is like.
5. Describe how people use air.
6. Tell how people use land and water.

Review Science Words

Match the words and the pictures.

1. stream
2. valley
3. mountain
4. soil

a.

b.

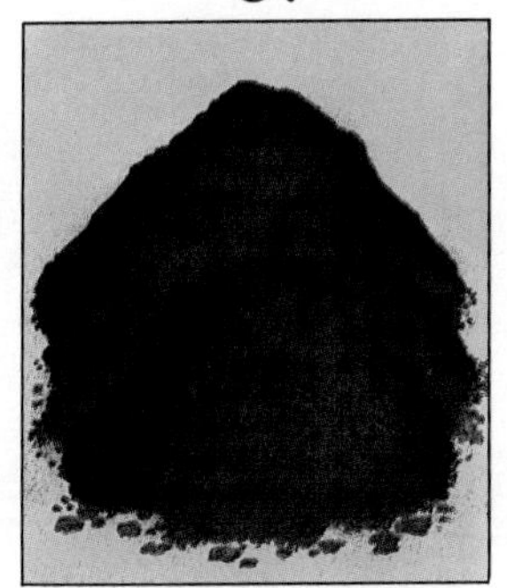

c.

d.

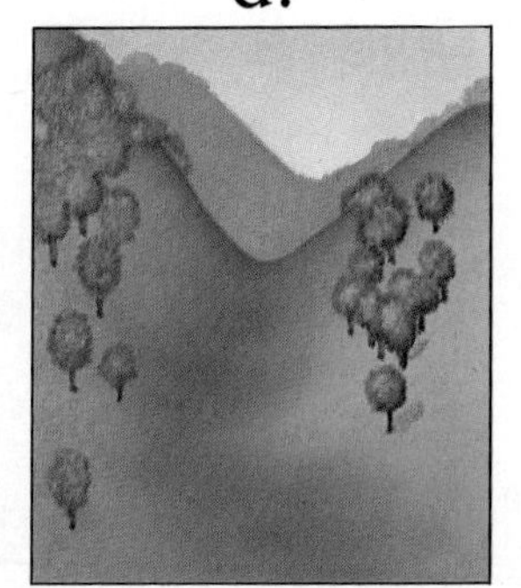

Tell what the words mean.

5. oil
6. coal
7. ocean

Use Science Ideas

Tell who is helping the earth.

Chapter 10

Weather and Seasons

Suppose the sun shines on these branches.
It gets warmer outside.
What might happen to the ice?

Starting the Chapter

You probably like to play outside.

You can observe when you are outside.

Do you notice if it is a warm or cold day?

Read how each day can be different.

TRY THIS

Observing the Weather

Look out the windows in your room.
Notice if the sky is clear or cloudy.
Notice if it is raining or snowing.
Draw a picture.
Show what the day is like.

Lesson 1 What Are Different Kinds of Weather?

Weather is what it is like outside.
Air temperature is part of weather.
The air can feel warm, hot, or cold.
It might snow on cold days.
Look at the picture.
What could you do on this day?

The weather can be cloudy or clear.
Clear weather has no clouds or rain.
The weather can be windy, too.
Clouds move with the wind.
Have you ever seen clouds like these?
This kind of cloud brings rain.

What weather can you see and feel?
You can see all kinds of weather.
You can feel wind, snow, and rain.
What kind of weather do you see here?

Lesson Review

1. What is weather?
2. What are five kinds of weather?

LIFE SCIENCE Find Out CONNECTION

How do animals keep safe in a storm?
Look in a book about animals.

Showing Air Temperature

Follow the Directions

1. Make three thermometer sticks.
2. Color in the sticks like this.
3. Glue the sticks to paper.
4. Mark the air temperature.
5. Write in the words.

Tell What You Learned

Tell the air temperature for each word.

Describe the weather for each temperature.

Lesson 2 How Can Weather Change in Seasons?

Many places on Earth have four **seasons.**
One season is **spring.**
How does the weather change in spring?
The air becomes warmer.
Some places have more rain.
Spring weather helps these plants grow.

Another season is **summer.**

The weather gets much warmer in summer.

The sun shines on many summer days.

Some places have rain in summer.

This place stays very dry.

What do you like to do in summer?

The weather changes in **fall,** too.
The air gets cooler.
Some places might have less rain.
The leaves of plants can change in fall.
What happens to these leaves?

Winter is the coldest season.
Some places have ice and snow.
Many plants do not grow in winter.
What is this winter day like?

Lesson Review

1. What are the four seasons?
2. How is weather different in seasons?

What is spring like in your area?
What plants do you see growing?
How does the weather change?

LIFE SCIENCE *Find Out* CONNECTION

Lesson 3 How Is Weather Important to People?

Weather is important for some jobs.
Pilots cannot fly in some weather.
Farmers need sunshine and rain.
Some farmers grow grapefruits.
They need warm weather to grow.
What protects these plants from cold?

Weather changes what people do.

What do people do in different weather?

Weather changes what people wear.

When would you use an umbrella?

Lesson Review

1. How is weather important for jobs?
2. What changes with weather?

EARTH SCIENCE

Find Out

Talk to your friends or family.

What kind of weather do they like?

Making a Weather Chart

Follow the Directions

1. Make a weather chart.
2. Show the weather for five days.
3. Draw one picture for each day.
4. Tell if it is cool, warm, or hot.

Tell What You Learned

Tell what your weather chart shows.

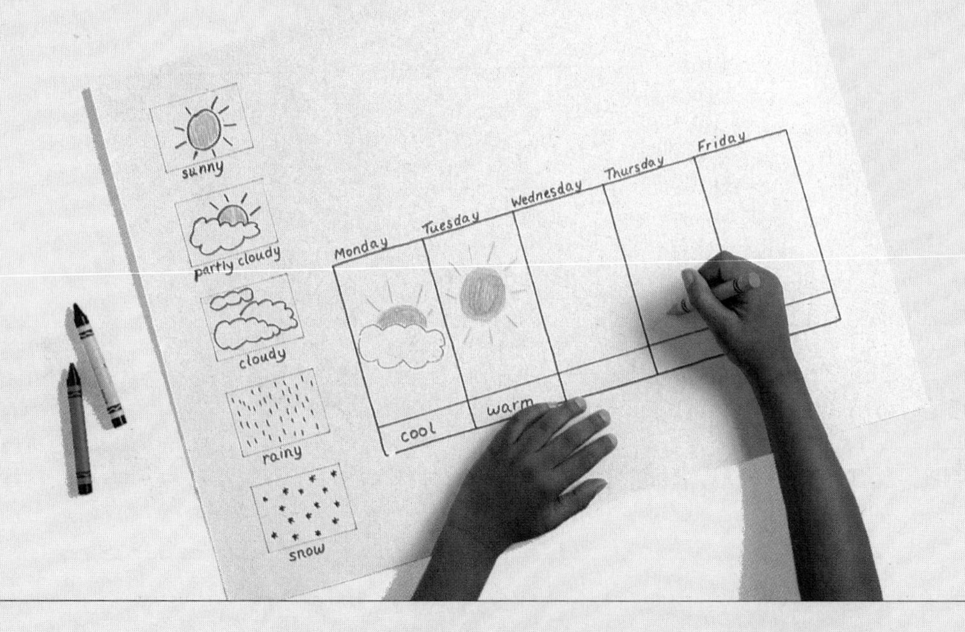

Taking Weather Pictures

Some spacecraft take weather pictures.

They travel far above the earth.

The pictures are sent to a computer.

People study the pictures.

They learn what weather is coming.

They let others know about the weather.

What Do You Think?

How do weather pictures help people?

Measuring Rain

How can you measure rain?

1. Look at the picture.

   It shows a way to measure rain.

   Notice how much it rained each day.

2. Make a chart like this one.
   Show how much rain fell each day.
   Find which days the most rain fell.
   Find which day the least rain fell.

Monday			
Tuesday	O	G	
Wednesday	O	G	R
Thursday	O	G	R
Friday	O		

3. What does measuring rain show?

Chapter 10 Review

Review Chapter Ideas

1. Describe different kinds of weather.
2. Name the four seasons.
3. Tell how weather changes in seasons.
4. Look at these people.
   Tell what kind of weather they need.

a.

b.

5. Tell how weather changes what people do.
6. Tell how weather changes what people wear.

Review Science Words

Match the words and the pictures.

1. fall
2. spring
3. summer
4. winter

a.

b.

c.

d.

Tell what the words mean.

5. weather
6. air temperature
7. seasons

a.

b.

Use Science Ideas

Look at the pictures.

Tell what the people would wear and do.

Chapter 11

The Sky

Suppose you watch the sky.
When does the sky look like this?

Starting the Chapter

What does the day sky look like?
What does the night sky look like?
Think of how it looks at day and night.
Then learn more about the sky.

TRY THIS

Looking at the Sky

Fold your paper into two parts.
Write the words you see here.
Draw the objects you see each time.
Tell how the sky looks different.

Lesson 1 What Do You See in the Sky?

You can see the sun in the sky.
You can see the **moon** and **stars.**
Pretend you could stand on the moon.
You would see the earth in the sky.
This picture shows how it might look.

Think about the sun, the earth, and the moon.
How are their sizes different?
The sun is much larger than the earth.
The earth is larger than the moon.

Lesson Review

1. What can be seen in the sky?
2. Is the earth, moon, or sun largest?

EARTH SCIENCE
Find Out

Go outside on a clear night.
Tell what you see in the sky.
Draw a picture to show it.

Lesson 2 What Is the Sun Like?

The sun is a hot ball of gas.
It is very, very large.
It is hotter than anything on the earth.
The sun heats the earth and moon.
The sun lights the earth and moon, too.
This boy is playing in the **sunlight.**

The earth is always turning.
Part of the earth is turned toward the sun.
The sun lights this part.
Sunlight makes **daytime** on the earth.
It is daytime on this part of the earth.

Part of the earth is away from the sun.
The sun does not light that part.
What does this part of the earth have?
This part of the earth has **nighttime.**

Lesson Review

1. What does the earth get from the sun?
2. What makes night and day?

EARTH SCIENCE
Find Out

Listen to a weather report.
Find out when the sun will set.

Ellison Onizuka

An astronaut flies in a spacecraft.
Ellison Onizuka was an astronaut.
He always liked to study the stars.
He worked hard to be an astronaut.
He flew in one of the Space Shuttles.
It went around the earth 48 times.

What Do You Think?

How do astronauts help others?

Lesson 3 What Is the Moon Like?

The moon is round like a ball.
It has rocks, soil, and mountains.
The moon has no light of its own.
What lights the moon?
Light from the sun shines on it.
Then the light bounces off the moon.

What part of the moon do you see?
You see the part lighted by the sun.
Sometimes you see the whole moon.
Other times you see part of the moon.
What shapes of the moon do you see here?

How do the moon and earth move?
The moon moves around the earth.
Together, they move around the sun.
Point to the sun, moon, and earth.

Lesson Review

1. What lights the moon?
2. In what ways does the moon move?

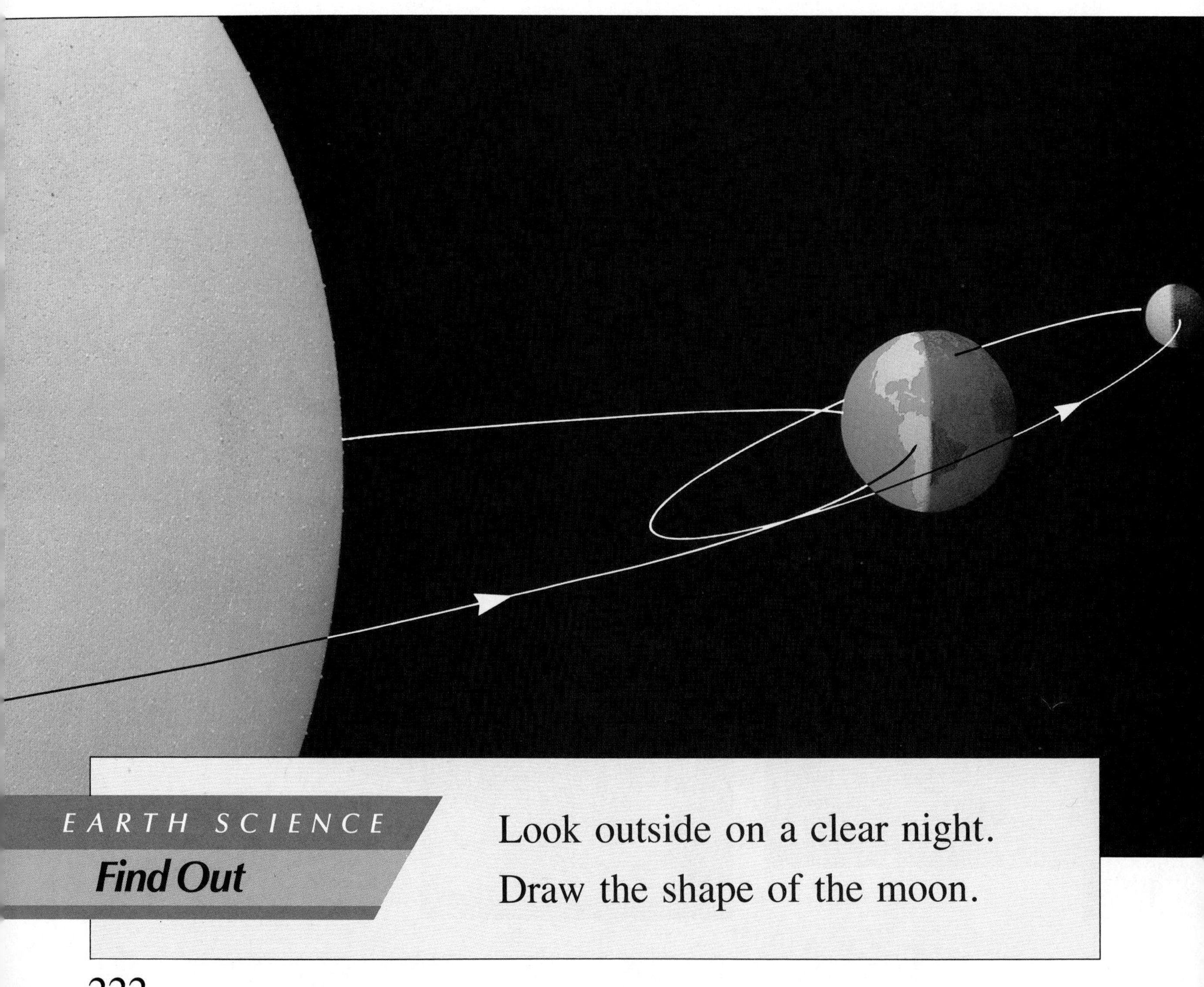

EARTH SCIENCE

Find Out

Look outside on a clear night.
Draw the shape of the moon.

Showing Day and Night

Follow the Directions

1. Use a foam ball to stand for Earth.
2. Put a paper person on the ball.
3. Have a partner shine a light this way.
4. Turn the ball on the stick.
5. Show the paper person in day and night.

Tell What You Learned

Show how the earth has day.

Show how the earth has night.

Lesson 4 What Are the Stars Like?

Stars are like the sun.
They shine with their own light.
Some stars are larger than the sun.
Stars are farther away than the sun.
They are so far away they look tiny.
When have you seen stars like these?

Many people study the stars.
They imagine pictures in the sky.
Stars can seem to make pictures.
This picture is the Big Dipper.
What other star pictures can you imagine?

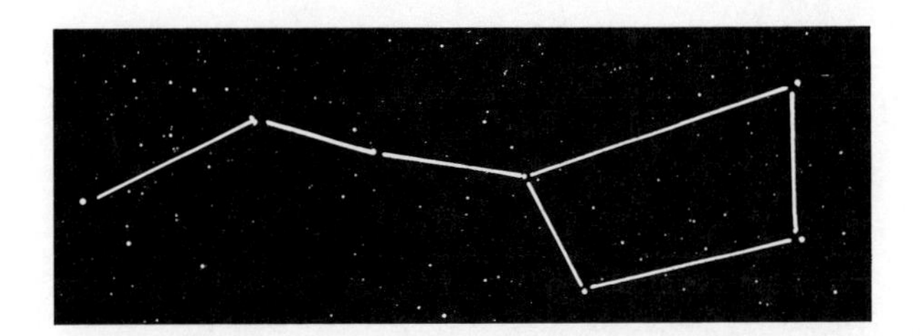

You cannot see most of the stars.
Most stars are too far away.
How can you see more stars?
You can use a **telescope** like this.

Lesson Review

1. How is a star like the sun?
2. Why do stars look small?

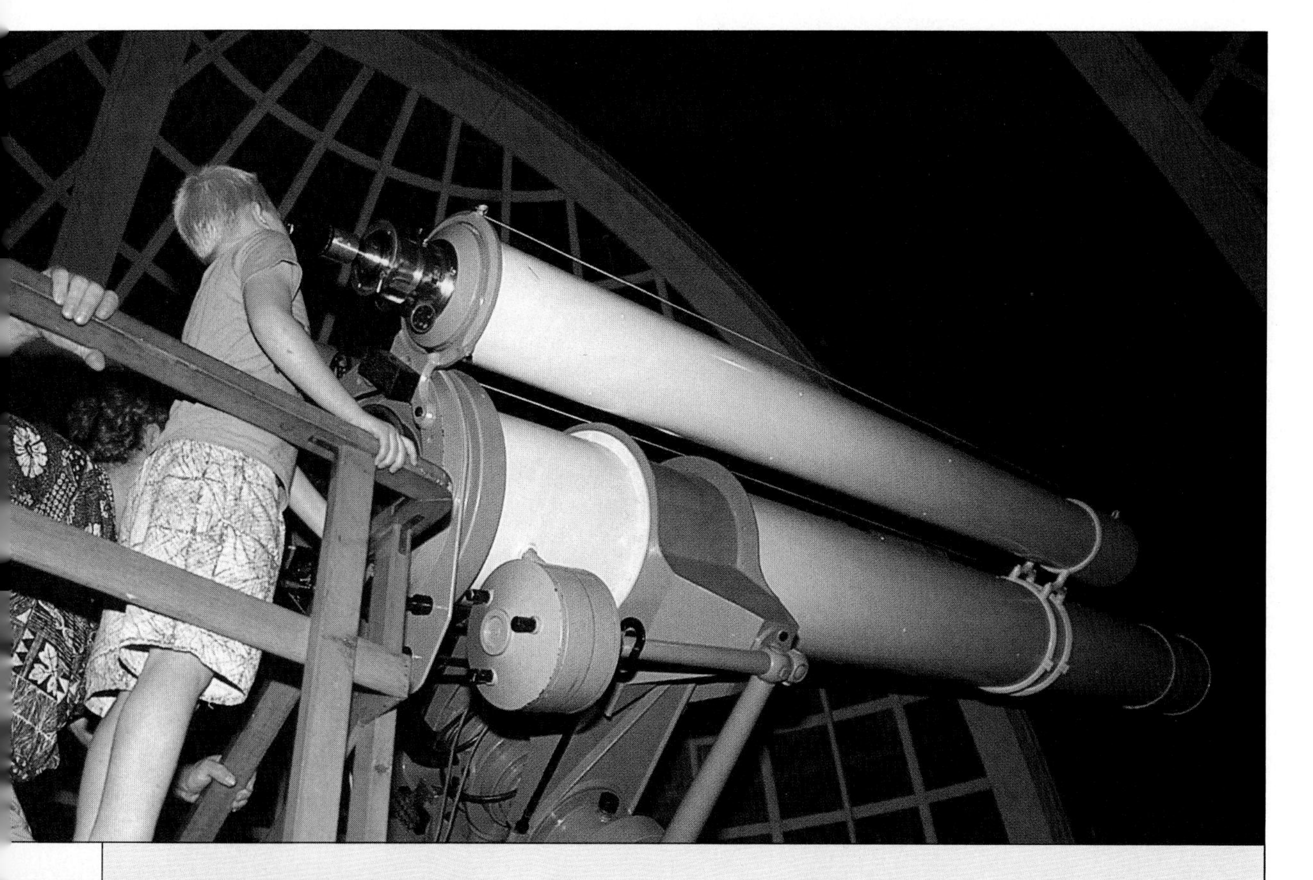

PHYSICAL SCIENCE
Find Out
CONNECTION

Telescopes make stars look larger.
Look in a book about stars.
Find more pictures of telescopes.

Making a Star Picture

Follow the Direction

1. Look at this star picture.
2. Copy it on your own paper.
3. Make a hole for each star.
4. Hold the paper up to a light.
5. Look at your star picture.

Tell What You Learned

Tell what your star picture shows.

Make a different star picture.

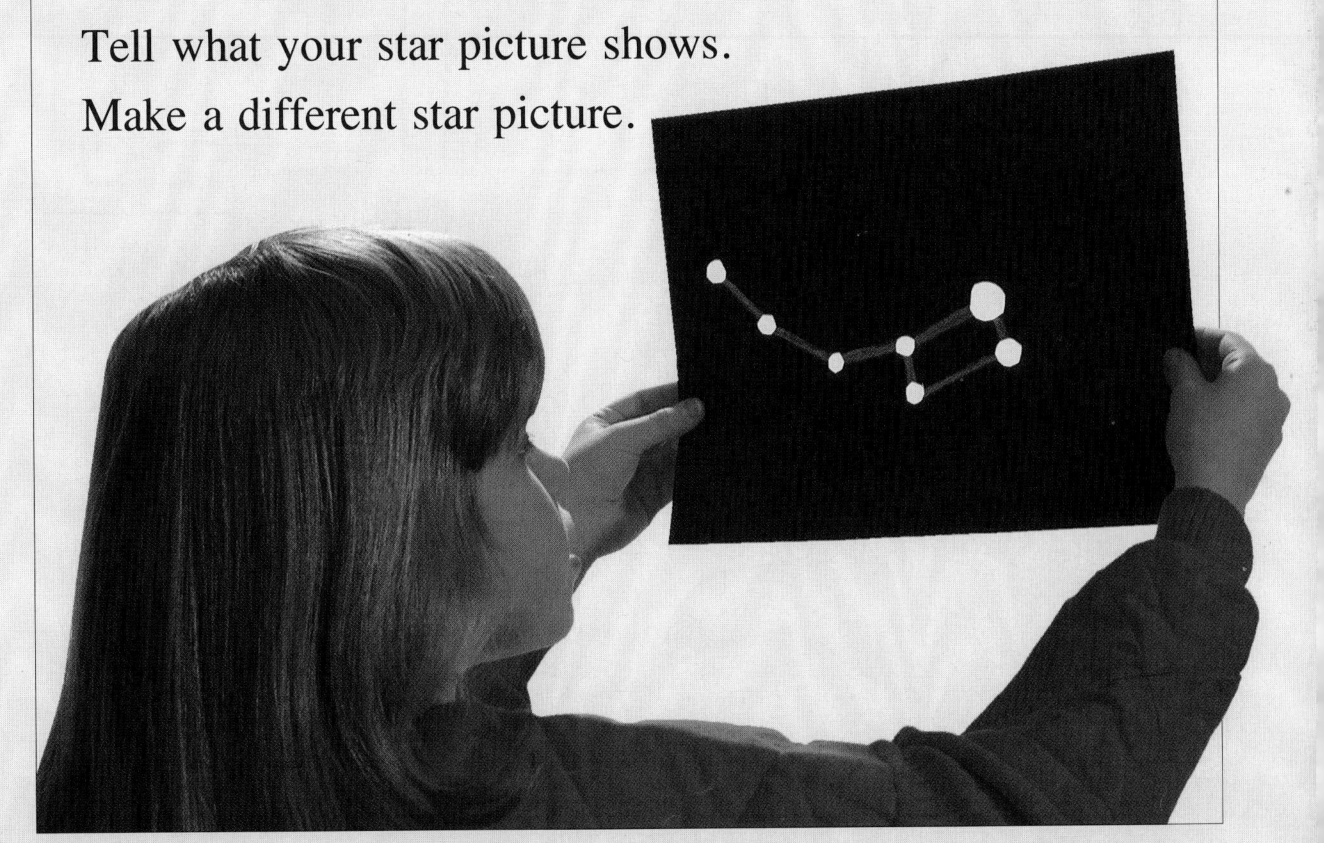

Measuring Shadows

How does a shadow change?

1. Look at the pictures.
   They show different times of day.
   Notice how the shadow changed.

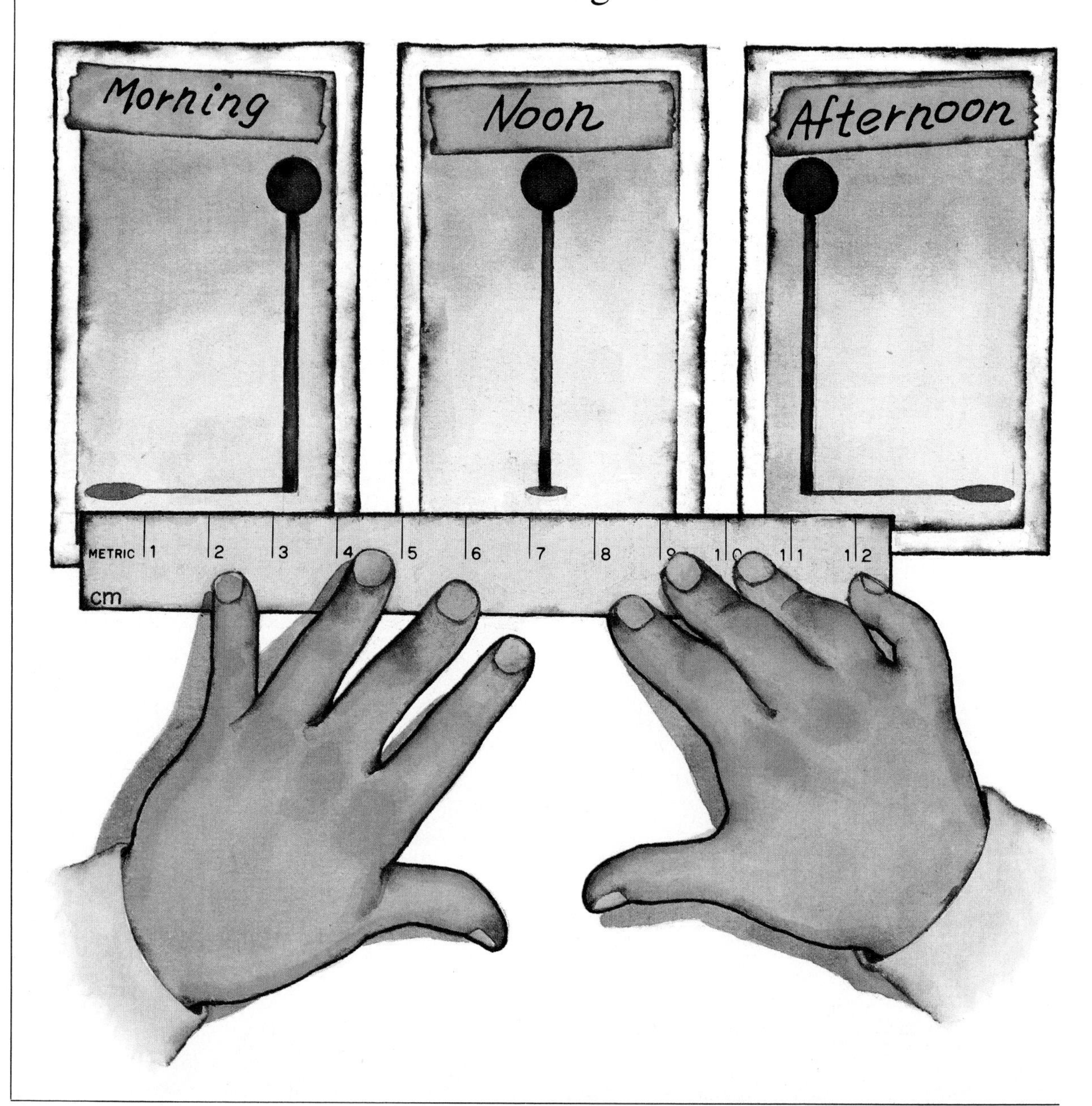

2. Measure each shadow.
   Use your own paper.
   Write how long each shadow is.
   Make a graph like this one.
   Fill in your graph.

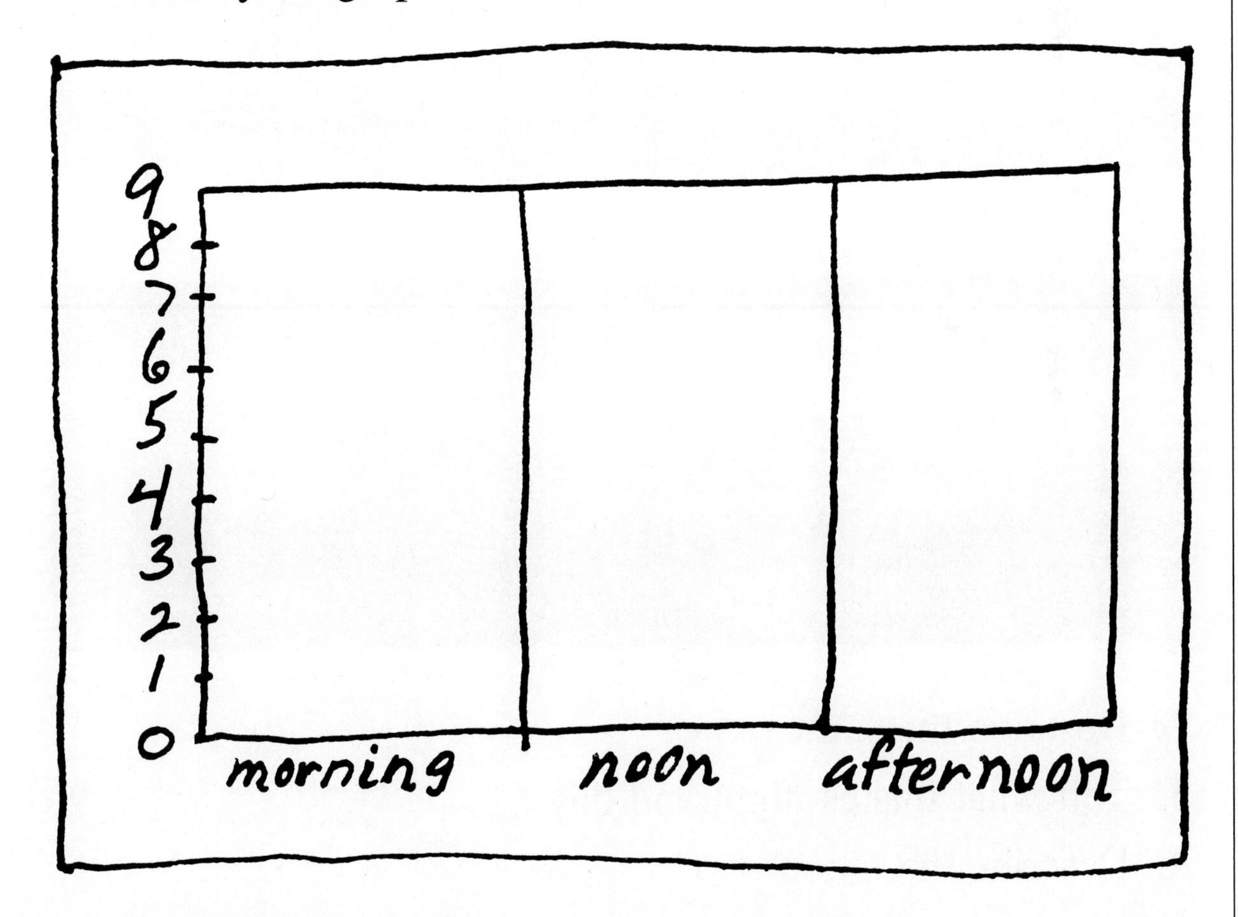

3. Look at your graph.
   How did the shadow change?

Chapter 11 Review

Review Chapter Ideas

1. Look at the pictures of the sky.
   Tell what you see.

a.

b.

c.

2. Describe the sun.
3. Tell what makes night and day.
4. Describe the moon.
5. Tell how the moon moves.
6. Describe the stars.

Review Science Words

Match the words and the pictures.

1. moon
2. sunlight
3. telescope
4. stars

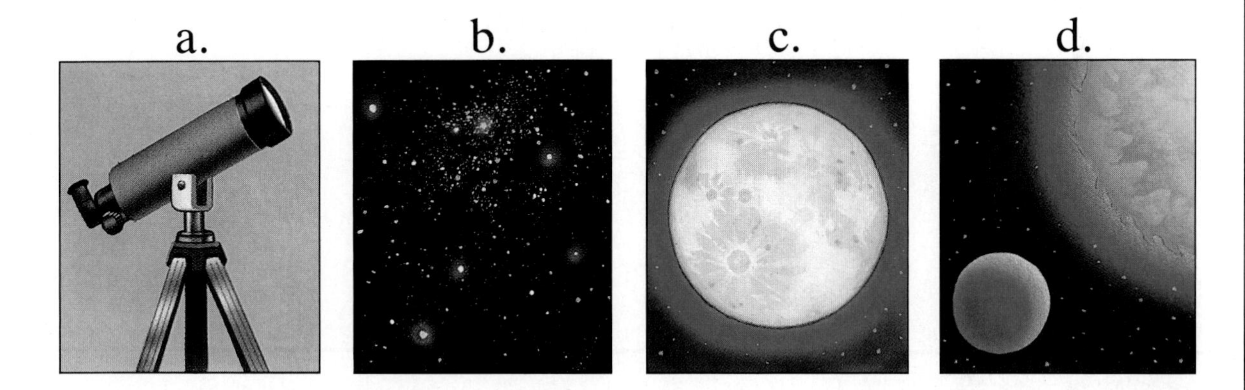

Tell what the words mean.

5. daytime
6. nighttime

Use Science Ideas

Which person will be in daytime first?

Scuba diver

Scuba divers can work under water.

They study living things in water.

Scuba divers take pictures under water.

Scuba divers carry tanks of air.

They breathe air from the tanks.

They can stay under water a long time.

Lawn Sprinkler

A sprinkler can spray water.
Water flows through the hose.
It quickly pushes into the sprinkler.
The water makes the sprinkler arms turn.
The sprinkler arms have small holes.
The water sprays out of the small holes.

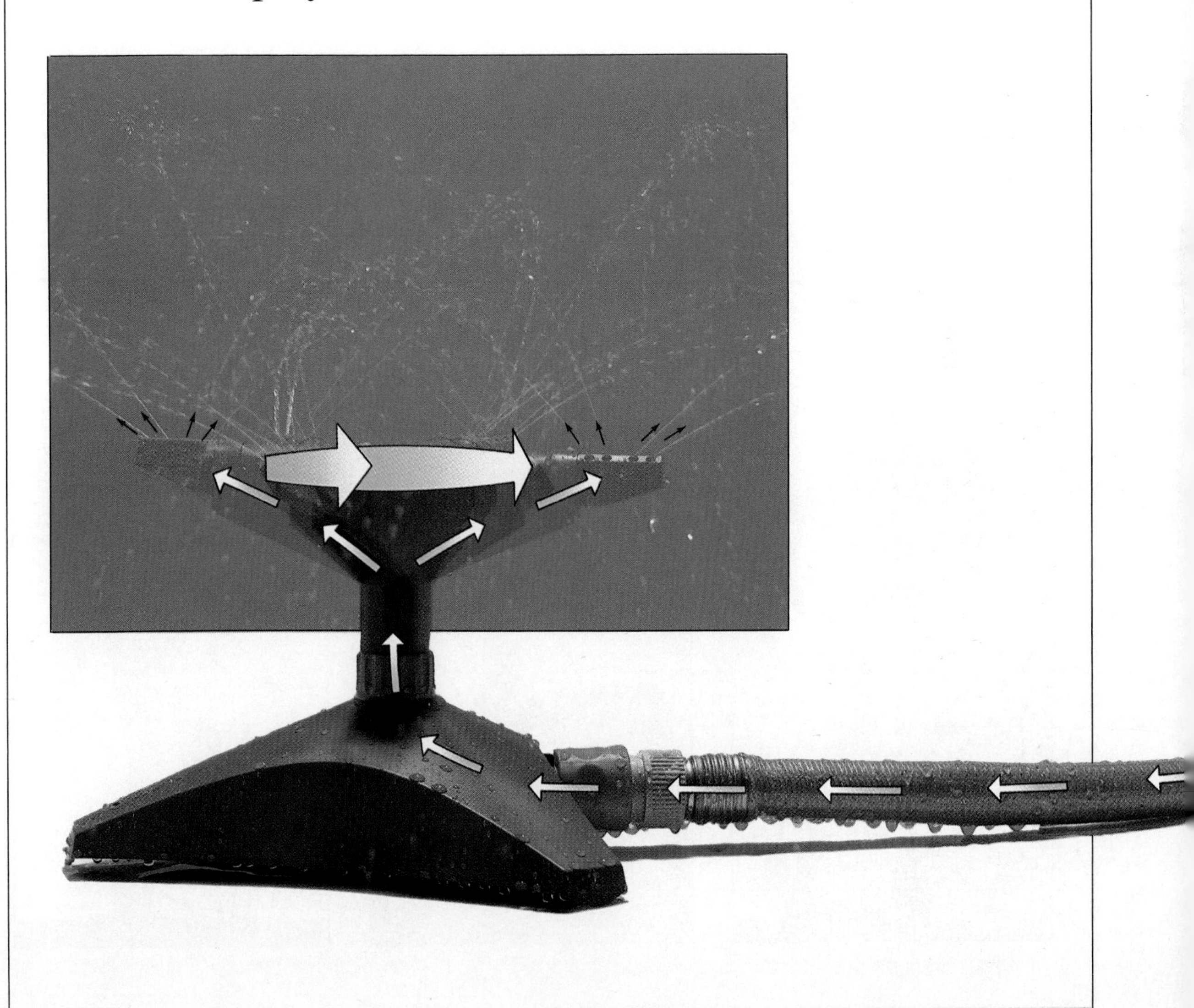

Unit 4 Review

Answer the Questions

1. What three things does the earth have?
2. How do people use air?
3. What are four kinds of weather?
4. How does weather change in seasons?
5. What are the sun and stars like?
6. How do the earth and moon move?

Study the Pictures

Look at the pictures.

When do plants grow best?

a.

b.

c.

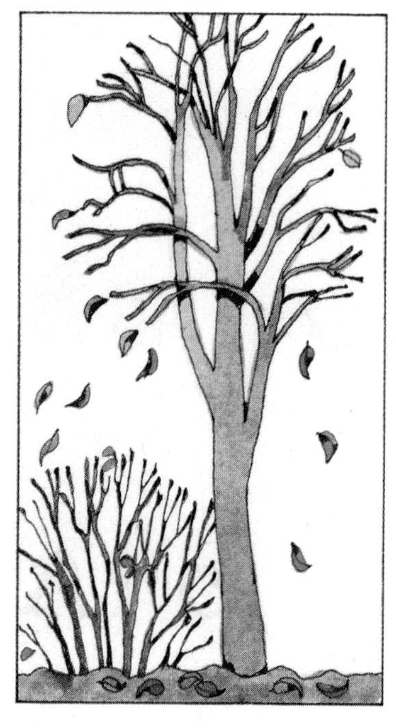

What to Do

1. Draw pictures of yourself.
   Show how you enjoy air, land, and water.
2. Answer this weather riddle.
   I help plants grow.
   I make puddles.
   I can come from clouds.
   What am I?
   Now make up your own weather riddle.
3. Pretend to take a trip into space.
   Make a model spacecraft.
   Use the kinds of things you see here.

Independent Study Guide

Answer the questions for each chapter.

Chapter 1 Study Guide

Use your own paper.

Write the best word.

LESSON 1
pages 16–18

1. What helps you hear?
   eyes ears nose
2. What helps you taste food?
   eyes ears tongue

LESSON 2
pages 20–23

3. What do you get from your senses?
   friends information food

Chapter 2 Study Guide

LESSON 1
pages 32–35

1. Your body ____ as you grow.
   looks changes tells
2. You get ____ as your bones grow.
   taller smaller friends
3. You get new teeth as you grow.
   The new teeth are ____ teeth.
   old smaller permanent

LESSON 2
pages 38–40

4. What is good exercise?
   writing reading playing
5. You need about ____ hours of sleep.
   two five ten
6. What helps you stay well?
   talking washing keeping

Chapter 3 Study Guide

LESSON 1
pages 54–56

1. Which are living things?
   books animals chairs
2. Many living things ____ on their own.
   ride sleep move
3. Living things can be ____.
   rocks parents water

LESSON 2
pages 58–60

4. Living things need ____.
   toys plants food
5. Ants find shelter in the ____.
   water ground plants

LESSON 3
pages 62–64

6. Look at the pictures
   Which is nonliving?

a.

b.

c.

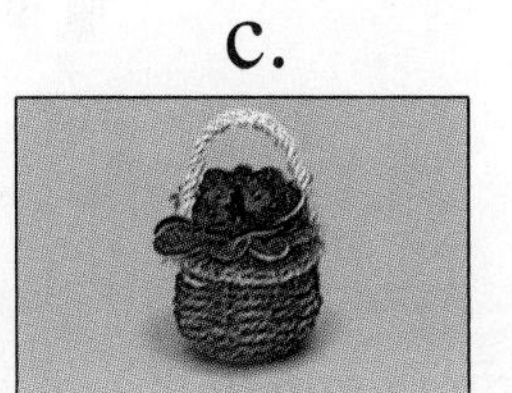

Chapter 4 Study Guide

Use your own paper.

Write the best word.

LESSON 1

pages 72–74

1. Plants have different shapes and ____.

   people colors pictures

2. Which is a part of plants?

   soil birds roots

LESSON 2

pages 76–78

3. Look at the picture.

   Which part makes seeds?

4. A seed ____ when it begins to grow.

   closes opens dies

LESSON 3

pages 80–82

5. What do plants get when it rains?

   air light water

6. What holds plants in place?

   air soil water

LESSON 4

pages 84–85

7. Paper comes from the ____ of trees.

   flowers seeds wood

Chapter 5 Study Guide

LESSON 1
pages 92–93

1. Coverings help ____ animals.
   hurt protect give
2. Which is a way some animals move?
   swim feel change

LESSON 2
pages 94–96

3. Many animals ____ when they grow.
   make use change

LESSON 3
pages 98–99

4. Which comes from animals?
   soil plants food

LESSON 4
pages 102–103

5. Which must people give their pets?
   clothing food shoes

Chapter 6 Study Guide

LESSON 1
pages 116–118

1. You can group objects by ____.
   buttons time color

LESSON 2
pages 120–121

2. A book and water both take up ____.
   time color space

LESSON 3
pages 122–123

3. Which has no shape of its own?
   milk book shoe

LESSON 4
pages 124–125

4. A gas can change ____ and shape.
   objects size liquids

Chapter 7 Study Guide

Use your own paper.

Write the best word.

LESSON 1

pages 134–136

1. Which makes light?

   chair window sun

2. Light can ▒▒ off a mirror.

   block bounce stop

LESSON 2

pages 138–140

3. Which can change from loud to soft?

   light smell sound

LESSON 3

pages 142–144

4. Fires give off ▒▒.

   heat air water

5. Which gives off heat?

   sun moon soil

Chapter 8 Study Guide

LESSON 1

pages 152–154

1. Look at the picture.

   The boy is ▒▒ the toy.

   above below near

LESSON 2

pages 156–158

2. Pushing and ▒ can move objects.

   sitting standing pulling

3. What do magnets do?

   play bounce pull

LESSON 3

pages 160–162

4. You ▒ to move a heavy box.

   read work sleep

5. What do people use for farming?

   paper magnets machines

Chapter 9 Study Guide

LESSON 1

pages 176–178

1. The earth is shaped like a ▒.

   box ball pen

2. What are most mountains made of?

   water sand rock

LESSON 2

pages 180–181

3. Most water on Earth is in ▒.

   rain oceans streams

LESSON 3

pages 182–184

4. People need air to ▒.

   smell taste breathe

LESSON 4

pages 186–188

5. People pump ▒ from the land.

   oil food air

Chapter 10 Study Guide

Use your own paper.

Write the best word.

LESSON 1
pages 196–198

1. Which can bring rain?
   snow   sun   clouds

LESSON 2
pages 200–203

2. Which season comes after spring?
   winter   summer   fall

LESSON 3
pages 204–205

3. Oranges need ____ weather to grow.
   warm   cold   winter

Chapter 11 Study Guide

LESSON 1
pages 214–215

1. The ____ is larger than the earth.
   sun   moon   cloud

LESSON 2
pages 216–218

2. Part of the earth faces the sun.
   This part of the earth has ____.
   nighttime   daytime   rain

LESSON 3
pages 220–222

3. The moon has ____ light of its own.
   much   some   no

LESSON 4
pages 224–226

4. Which can you use to see more stars?
   sunlight   weather   telescope

Using Scientific Methods

Scientists like to learn about the world.
They like to help with problems.
They use scientific methods to find answers.
Scientists use many steps in their methods.
Sometimes they use the steps in different ways.
You can use these steps to do experiments.

Explain the Problem

Ask a question like this.
Do plants grow toward light?

Make Observations

Tell about the size, color, or shape of something.

Give a Hypothesis

Try to answer the problem.
Tell your idea.
Then do the experiment.

Make a Chart or Graph

Tell what you saw in your chart or graph.

Make Conclusions

Decide if your hypothesis is right or wrong.

Safety in Science

Scientists are careful when they do experiments.
You need to be careful too.
Here are some rules to remember.

- Read each experiment carefully.
- Wear cover goggles when needed.
- Clean up spills right away.
- Never taste or smell unknown things.
- Do not shine lights in someone's eyes.

- Put things away when you are done.
- Wash your hands after each experiment.

Chapter 1 Experiment Skills

Matt just became a crossing guard.
He brought home his orange belt.
His sister Ellen thinks orange is ugly.
She likes blue.
Ellen wonders why the belts are orange or yellow.
She notices that the school bus also is yellow.
She wonders if yellow is easy to see.

Problem

Is it easier to see yellow than another color?
Give your hypothesis.
Then find out if your hypothesis is right.
Read the experiment to find out.

Follow the Directions

1. Make a chart like the one below.
2. Get a piece of bright yellow paper.
   Tape it against a gray paper.
   Can you see the yellow easily?
   Write your answer in your chart.
3. Then get a piece of blue paper.
   Tape it against the gray paper.
   Can you see the blue easily?
   Fill in your chart.
4. Which color is easier to see?
   Circle the color in your chart.

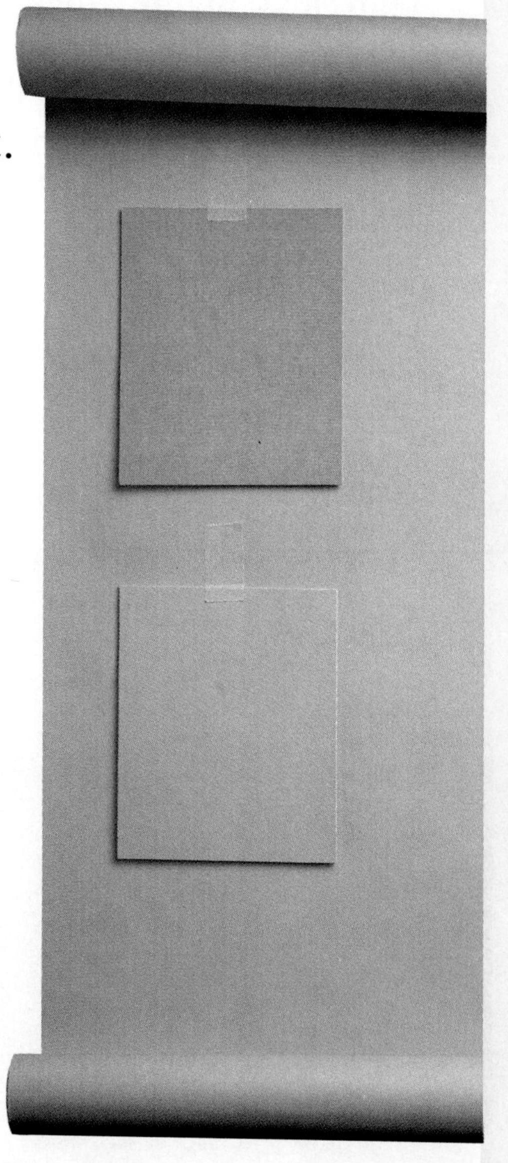

Tell Your Conclusion

Is it easier to see yellow than another color?

Color	Easier to see
yellow	
blue	

Chapter 2 Experiment Skills

Jack rinses his mouth after lunch.
This makes his mouth feel clean.
He wonders why he needs to brush.
Can just water clean his teeth?
Jack wonders if brushing cleans better.

Problem

Does brushing help make things clean?
Give your hypothesis.
Then find out if your hypothesis is right.
Read the experiment to find out.

Follow the Directions

1. Make a chart like the one below.
2. Rub peanut butter on two spoons.
   Let the spoons sit for two hours.
3. Run water over one spoon.
   Does rinsing help clean the spoon?
4. Scrub the other spoon with a brush.
5. Does brushing help clean the spoon?
   Write your answers in your chart.
6. Circle in the chart which way cleans better.

Tell Your Conclusion

Does brushing help make things clean?

What helps clean	Is spoon clean?
rinsing	
brushing	

Chapter 3 Experiment Skills

Lisa and her grandfather went to a pet shop.
She saw many things that looked like fish in a tank.
Her grandfather said that they were not fish.
They were tadpoles.
Lisa went back to the pet shop many weeks later.
The tank had many frogs.
Lisa wonders if the tadpoles changed into frogs.
She wonders if other animals change too.

Problem

Do some animals change color and shape?
Give your hypothesis.
Then find out if your hypothesis is right.
Read the experiment to find out.

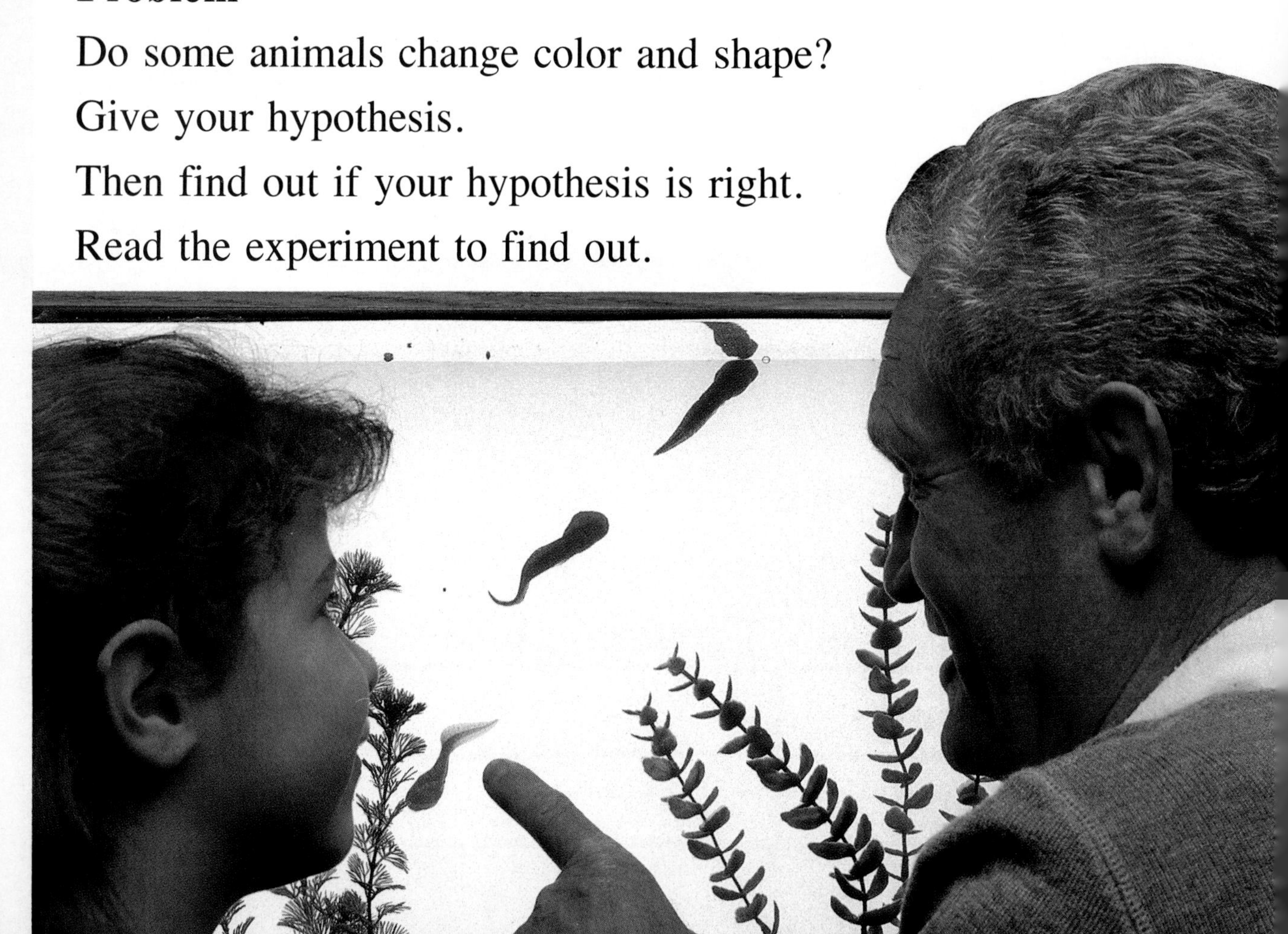

Follow the Directions

1. Make a chart like the one below.
2. Put some mealworms in a jar.
   What color and shape are the animals?
   Write your answer in your chart.
3. Put cereal and apple pieces in the jar.
   Cover it with cheesecloth.
4. Look at the animals every week for a month.
   Feed them once a week.
5. How do the animals look after one month?
   Write in your chart how they look.

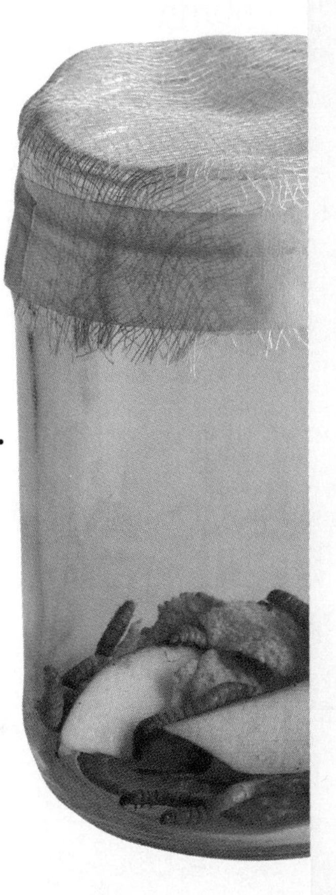

mealworm

beetle

Tell Your Conclusion

Do some animals change shape and color?

Animal	Color and shape
mealworm	
beetle	

Chapter 4 Experiment Skills

Brad is watering his friend's plants.
One plant is bent.
It is leaning toward the window.
Brad thinks the plant is bending toward light.
He wonders if other plants bend toward light.

Problem

Do some plants bend toward light?
Give your hypothesis.
Then find out if your hypothesis is right.
Read the experiment to find out.

Follow the Directions

1. Make a chart like the one below.
2. Put a small plant near a sunny window for one week.
3. Does the plant bend toward light?
   Write yes or no in your chart.
4. Turn the plant so it bends away from the window.
   Leave it for one week.
5. Write in your chart if the plant bends now.

after one week

turn plant

one week later

Write Your Conclusion

Do some plants bend toward light?

Time	Bends toward light
after 1 week	
1 week after turning plant	

Chapter 5 Experiment Skills

Joe was helping his mother clean.
His mother took a cover off a light.
The cover had many flies.
Joe wondered where the flies came from.
He thinks the flies flew to the light.

Problem

Do flies go toward light?
Give your hypothesis.
Then find out if your hypothesis is right.
Read the experiment to find out.

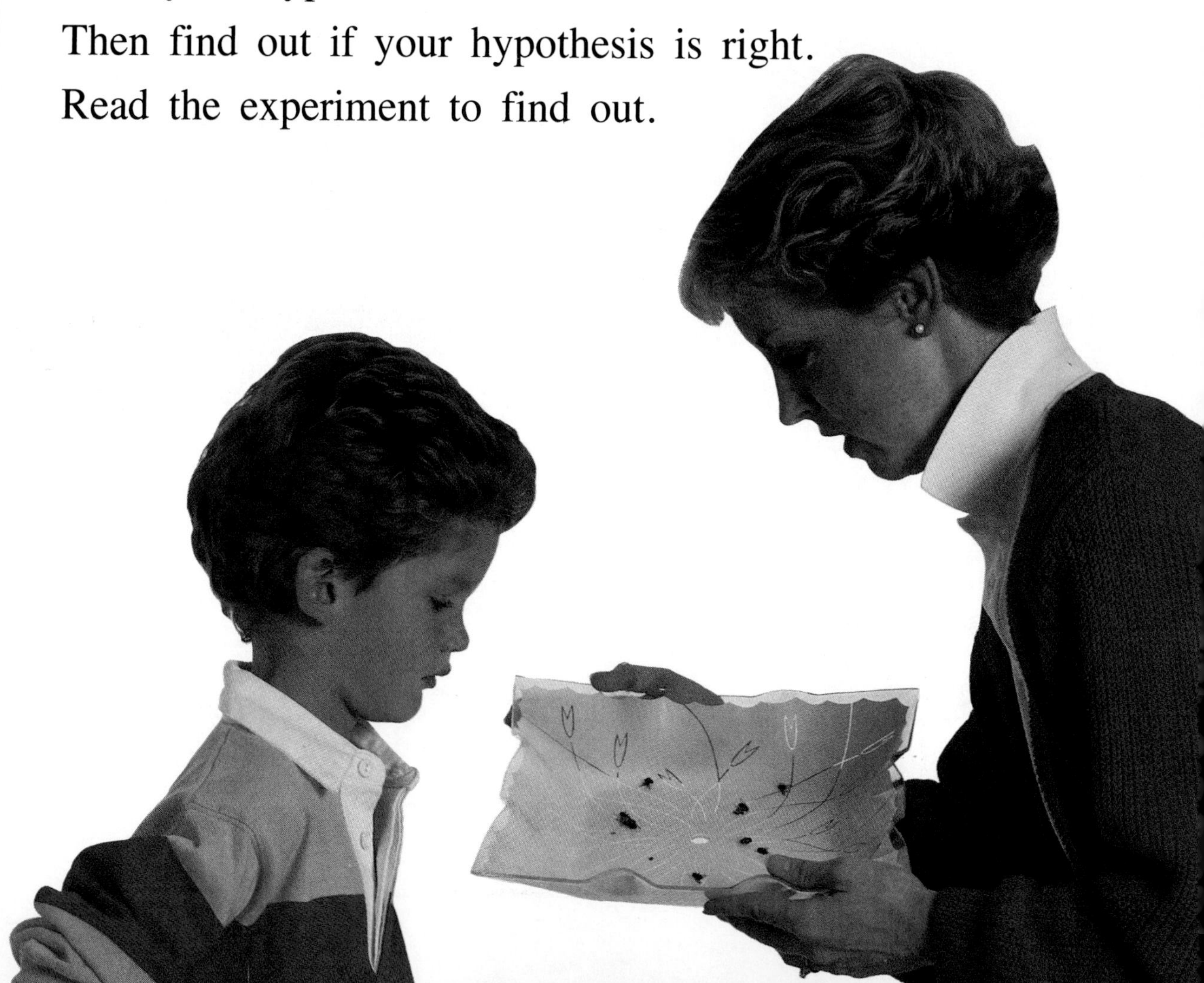

Follow the Directions

1. Make a chart like the one below.
2. Put six fruit flies in a jar.
   Cover the jar with cheesecloth.
   Tape black paper around the bottom.
3. Shine a flashlight at the jar.
4. Do the flies go to the light?
   Do the flies go to the dark?
   Tell your answers in your chart.

Tell Your Conclusion

Do flies go toward light?

Part of jar	Flies go
light	
dark	

Chapter 6 Experiment Skills

Beth and Tim like to play in the snow.
They made some snowmen in the morning.
The snowmen were gone by afternoon.
Only a pile of snow and some water were left.
Tim thinks warm air melted the snowmen.

Problem

Does warm air change snow and ice into a liquid?
Give your hypothesis.
Then find out if your hypothesis is right.
Read the experiment to find out.

Follow the Directions

1. Make a chart like the one below.
2. Get two ice cubes from the freezer.
3. Put each ice cube in a paper or plastic cup.
4. Put one cup in a cold place.
   Put the other cup in a warm place.
5. Wait one hour.
6. Which cup has a solid?
   Which cup has a liquid?
   Write your answers in your chart.
   Circle in the chart where the ice changes shape.

Tell Your Conclusion

Does warm air change snow and ice into a liquid?

Place	Object in cup
cold place	
warm place	

Chapter 7 Experiment Skills

Kim sees a circle of light on the wall.
She wonders where the light comes from.
Kim notices a shiny tray on the table.
Light is hitting the tray.
She thinks light is bouncing off the tray.
Then the light hits the wall.

Problem

Does light bounce off shiny objects?
Give your hypothesis.
Then find out if your hypothesis is right.
Read the experiment to find out.

Follow the Directions

1. Make a chart like the one below.
2. Shine a light toward a mirror.
   Move the mirror around a little.
   Do you see a circle of light on the wall?
3. Cover the mirror with black paper.
   Hold it toward the light.
   Do you see a circle of light on the wall?
4. Write your answers in your chart.

Tell Your Conclusion

Does light bounce off shiny objects?

Object	Light bounces
shiny	
dark	

Chapter 8 Experiment Skills

Mary spilled paper clips all over the table.
Bob offers to help her pick them up.
Mary thinks they should use magnets.
Bob wants to use two magnets together.
He thinks two magnets will make the job easier.

Problem

Are two magnets stronger than one?
Give your hypothesis.
Then find out if your hypothesis is right.
Read the experiment to find out?

Follow the Directions

1. Make a chart like the one below.
2. Dump out a box of paper clips.
3. Use one magnet to pick up the clips.
4. How many did you pick up?
   Write the number in your chart.
5. Use two magnets to pick up clips.
6. How many did you pick up?
   Write the number in your chart.

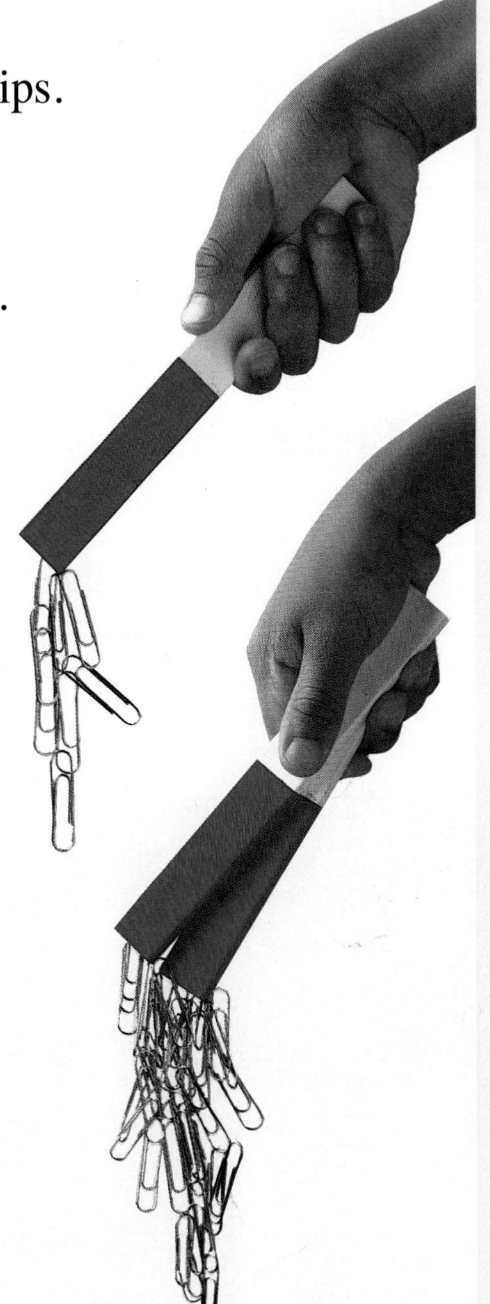

Tell Your Conclusion

Are two magnets stronger than one?

Number of Magnets	Number of Clips
1	
2	

Chapter 9 Experiment Skills

Sally went out early today.
She saw water on the grass.
It did not rain during the night.
But it was cool.
Sally wonders about the water.
She thinks it came out of the cool air.
She wants to find out if air has water.

Problem

Does air have water in it?
Give your hypothesis.
Then find out if your hypothesis is right.
Read the experiment to find out.

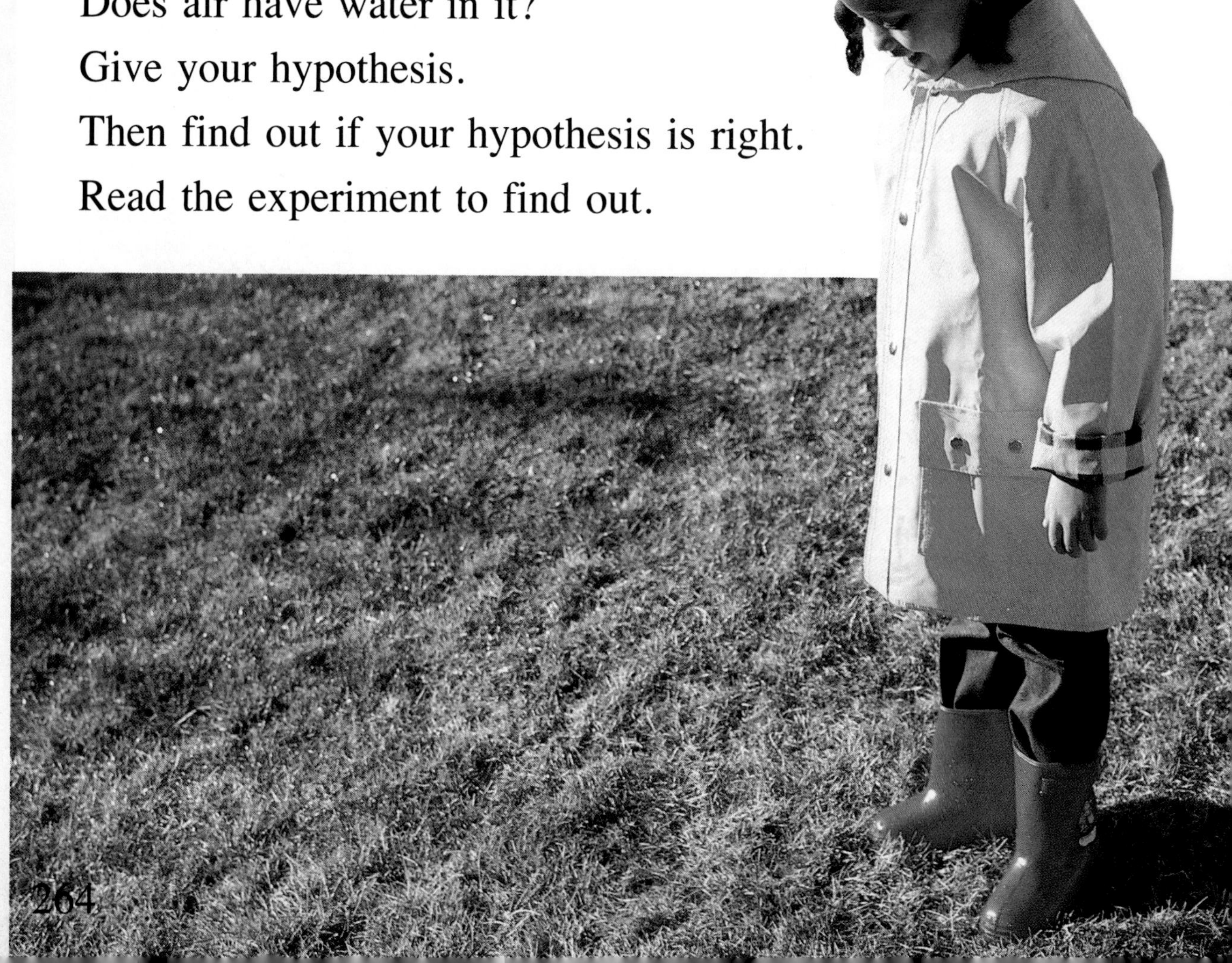

Follow the Directions

1. Make a chart like the one below.
2. Rub your finger across a mirror.
   Tell in your chart if your finger is wet.
3. Put the mirror in a cold place.
   Leave it there for 15 minutes.
4. Get the mirror.
   Blow air from your mouth on it.
5. Rub the mirror with your finger.
   Write in your chart if your finger is wet.

Tell Your Conclusion

Does air have water in it?

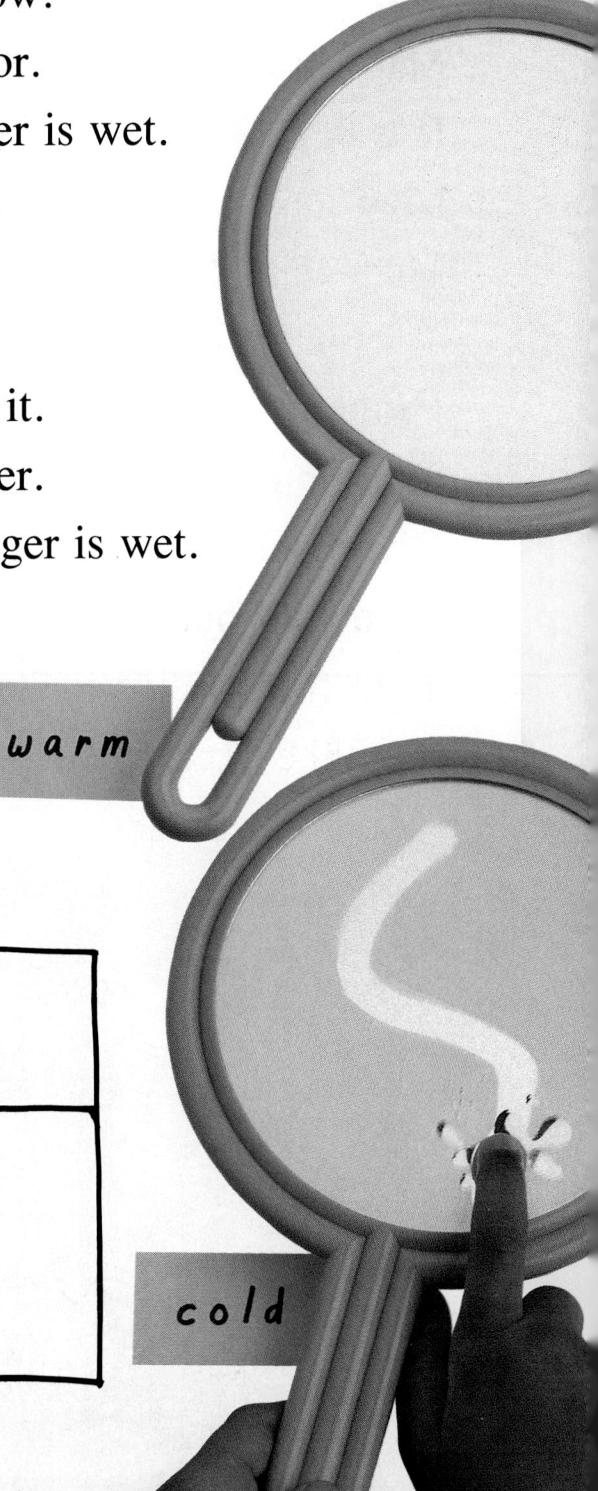

Mirror	Wet
warm	
cold and cloudy	

Chapter 10 Experiment Skills

Mary is walking home from school.
She takes off her coat.
Mary needed her coat this morning.
It was cool then.
Now it is warm.
Mary thinks the air gets warmer during the day.

Problem

Can air get warmer during the day?
Give your hypothesis.
Then find out if your hypothesis is right.
Read the experiment to find out.

Follow the Directions

1. Make a chart like the one below.
2. Put a thermometer outside.
3. Look at it in the morning.
   See how high the red line is.
4. Look at the thermometer in the afternoon.
   Did the red line go up?
5. Look at the thermometer for two days.
   Look in the morning and afternoon.
6. When is the line highest?
   When is the line lowest?
   Write your answers in your chart.
7. Circle which time of day is warmer.

morning

Tell Your Conclusion

Can air get warmer during the day?

Time	Thermometer line
morning	
afternoon	

afternoon

Chapter 11 Experiment Skills

It is a warm summer morning.
Dan wants to swim in his pool.
His mother says the water is too cold.
She tells Dan he can swim later.
She says the water will be warmer.
Dan wonders if the sun heats water.

Problem

Can the sun heat water?
Give your hypothesis.
Then find out if your hypothesis is right.
Read the experiment to find out.

Follow the Directions

1. Make a chart like the one below.
2. Fill two cups with cold water.
3. Put one cup in a sunny place.
4. Put the other cup in a shady place.
5. Wait for two hours.
6. Put a thermometer in each cup.
7. Which thermometer has a longer red line?
   Which thermometer has a shorter line?
   Write your answers in your chart.
8. Circle which cup has warmer water.

Tell Your Conclusion

Can the sun heat water?

Water	Thermometer line
cup in sun	
cup in shade	

Glossary/Index

A

air temperature, page 196. The *air temperature* is how warm or cold air feels.

B

below, page 152. This butterfly is *below* the flower.

bones, page 32. You have many *bones* in your body.

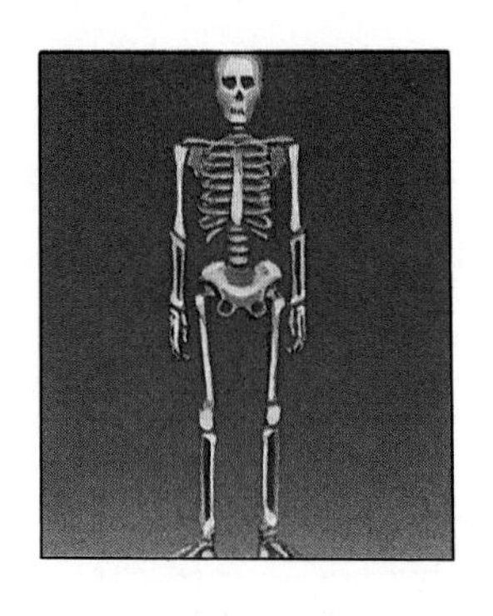

bounce, page 135. Light can *bounce* off a solid object.

C

coal, page 178. People can burn *coal* for heat.

coverings, page 92. Feathers, fur, and shells are some animal *coverings*.

D

daytime, page 217. It is *daytime* between sunrise and sunset.

direction, page 139. North, south, east, and west are each a *direction*

distance, page 153. The space between two things is *distance*.

E

electric light, page 134. Turn on an *electric light* to see better.

exercise, page 39. You can *exercise* to stay healthy.

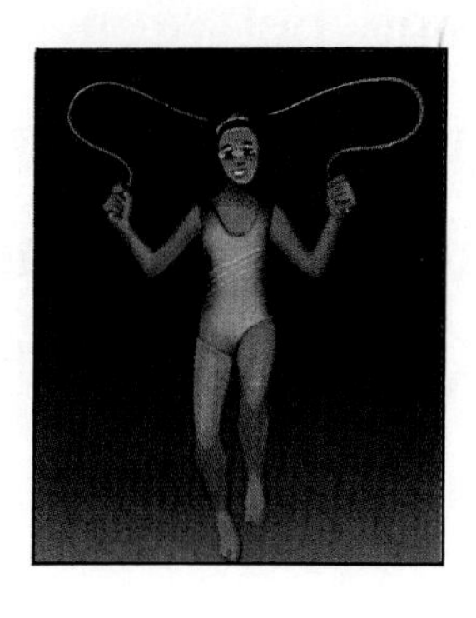

F

fall, page 202. The season between summer and winter is *fall*.

G

gas, page 121. The *gases* that fill a balloon are air.

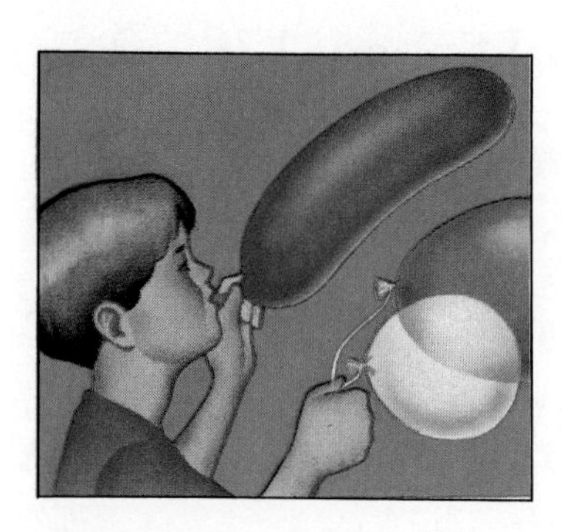

H

healthy, page 38. Eat well and exercise to stay *healthy*.

hearing, page 16. You listen with your sense of *hearing*.

heat, page 142. The sun's *heat* warms the earth.

hypothesis, page 3. A *hypothesis* is a possible answer to a problem.

I

insect, page 96. An *insect* is a kind of very small animal.

L

leaves, page 73. Roots, stems, and *leaves* are parts of plants.

liquid, page 120. A *liquid* changes shape as it is poured from a glass to a cup.

living thing, page 54. A plant and an animal are each a *living thing*.

M

machine, page 160. A *machine* can help you work.

magnet, page 158. A *magnet* can attract some objects.

moon, page 214. The *moon* is bright in the sky at night.

mountain, page 176. A *mountain* is a very high hill.

muscles, page 32. You use your *muscles* when you move.

N

nighttime, page 218. The time between sunset and sunrise is *nighttime*.

nonliving thing, page 62. A rock and a book are each a *nonliving thing*.

O

object, page 116. An *object* is something you can see.

observe, page 20. You use your senses to *observe* things.

ocean, page 180. An *ocean* is a very large body of salt water.

oil, page 178. People pump *oil* from the earth.

P

parent, page 56. A *parent* is a mother or a father.

permanent teeth, page 33. Your *permanent teeth* take the place of your baby teeth.

pet, page 102. A *pet* is an animal you take care of.

protect, page 92. An animal's coverings help *protect* it.

R

root, page 73. A *root* is a part of a plant.

S

season, page 200. Winter, spring, summer, and fall are each a *season*.

seed, page 76. If you plant a *seed* in the ground, it might grow.

seeing, page 16. When you are looking at something, you are *seeing* it.

senses, page 16. Your five *senses* are seeing, hearing, feeling, tasting, and smelling.

shadow, page 136. When something blocks the light, it makes a *shadow*.

shelter, page 59. A *shelter* can be something that covers or protects.

smelling, page 16. When you breathe in an odor, you are *smelling* it.

snake, page 96. A *snake* is a long, thin animal.

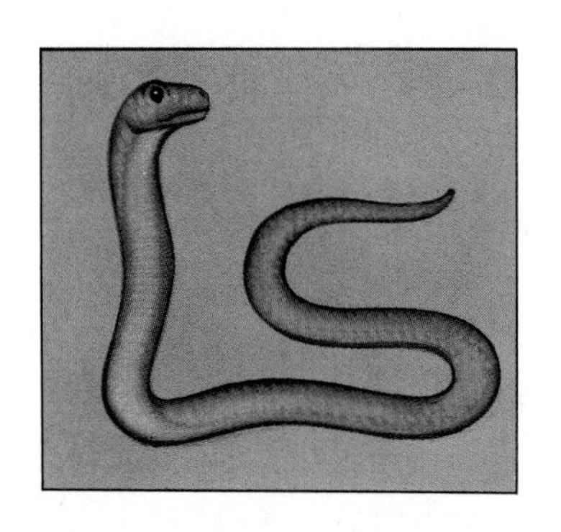

soil, page 81. Another word for dirt is *soil*.

solid, page 120. An object that is a *solid* takes up space and has its own shape.

space, page 120. The puzzle piece is the right shape to fill the *space*.

speed, page 154. How fast something moves is its *speed*.

spring, page 200. The season between winter and summer is *spring*.

star, page 214. You might see a *star* in the sky at night.

stem, page 74. A *stem* is part of a plant.

stream, page 180. A *stream* is running water.

summer, page 201. The season between spring and fall is *summer*.

sunlight, page 216. A plant can use *sunlight* to help it grow.

T

tasting, page 16. When you put something in your mouth, you are *tasting* it.

telescope, page 226. A *telescope* can help you see stars in the sky.

thermometer, page 144. A *thermometer* can tell how hot an object is.

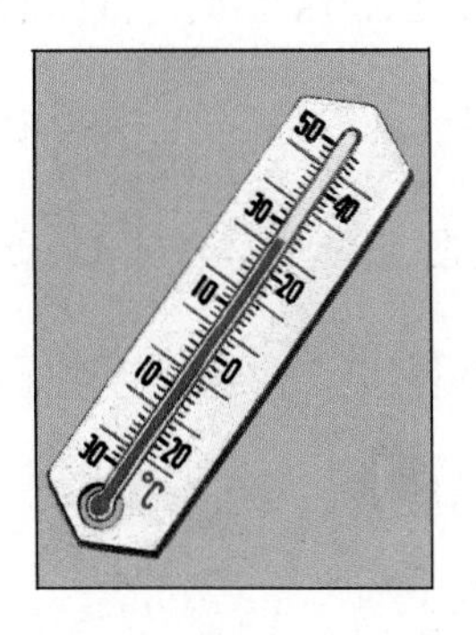

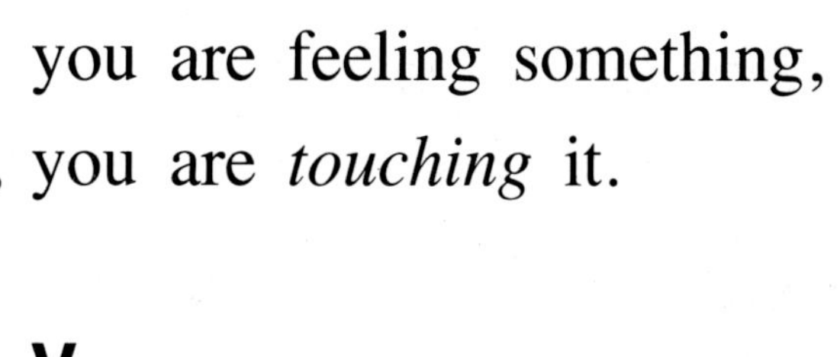

touching, page 16. When you are feeling something, you are *touching* it.

V

valley, page 176. A *valley* is a place between two mountains or hills.

W

weather, page 196. The *weather* today is sunny and cool.

winter, page 203. The season between fall and spring is *winter*.

Acknowledgments

Unless otherwise acknowledged, all photos are the property of Scott, Foresman & Company. Page positions are as follows: (T)top, (B)bottom, (C)center, (L)left, (R)right, (INS)inset.

Page **IV:** Howard Sochurek **VI:** Sullivan & Rogers/Bruce Coleman Inc. **VII(L):** Color Advantage/Robert E. Lyons **VII(R):** William E. Ferguson Photography **X:** NASA **XI:** David R. Frazier Photolibrary **XII:** Courtesy Shedd Aquarium **1:** Al Grotell **2-3:** Jeff Rotman/Peter Arnold, Inc. **4TL:** Fred Bavendam/Peter Arnold, Inc. **4TR:** Alex Kerstitch/Sea of Cortez Enterprises **4BL:** Fred Bavendam/Peter Arnold, Inc. **4BR:** Fred Bavendam/Peter Arnold, Inc. **5(TOP TO BOTTOM):** Dr. E. R. Degginger Dr. E. R. Degginger Hans Phetschinger/Peter Arnold, Inc. Dr. E. R. Degginger **6T:** Dr. E. R. Degginger **6B:** Phil Degginger **12:** (c) 1984/Howard Sochurek **14:** David Louis Olson **17R:** David Phillips **12BR:** Jose Azel/Contact Press Images/Woodfin Camp & Associates **20-21:** Harald Sund **30:** REPRINTED FROM PSYCHOLOGY TODAY MAGAZINE Copyright (c) 1987 **37:** Lewis Watts **50:** Kjell Sandved/Sandved and Coleman Photography **54:** Michael Fairchild/Peter Arnold, Inc. **55L:** Walter Chandoha **55R:** Walter Chandoha **56L:** G. Ziesler/Peter Arnold, Inc. **56R:** E.R. Degginger/Bruce Coleman Inc. **58-59:** Hans Reinhard/Bruce Coleman Inc. **59INS:** Wolfgang Bayer Productions **61:** NASA **66R:** Color Advantage/Robert E. Lyons **69C:** Marty Snyderman **70:** Mary E. Goljenboom/Ferret **72L:** Don and Pat Valenti **74L:** J. Serrao **74R:** Ira Cohen/New England Stock Photo **76:** Kjell Sandved/Sandved and Coleman Photography **78TL:** Don and Pat Valenti **78BC:** Lynn M. Stone **78R:** Stephenie S. Ferguson **79L:** Lynn M. Stone **80:** Lynn M. Stone **81:** Grant Heilman Photography **82:** J. Serrao **83:** William James/West Light **84:** Don and Pat Valenti **85L:** R. Hamilton Smith **85R:** Grant Heilman Photography **90:** Kjell Sandved/Sandved and Coleman Photography **92TL:** (c) 1983/Carl Roessler **92TR:** Bill Ivy **92BL:** Sullivan & Rogers/Bruce Coleman Inc. **92-93:** MPL Fogden/Bruce Coleman Inc. **93INS:** Lynn M. Stone **94(ALL):** William E. Ferguson Photography **95B:** Arthus-Bertrand/Peter Arnold, Inc. **95R:** Wayne Lankinen/DRK Photo **96L:** D. Wilder **96R:** Dr. E.R.Degginger/Bruce Coleman Inc. **99:** (c) Mickey Pfleger 1987 **101:** Robert Rattner **102-103:** Frank Popper/Photographic Resources, Inc. **103L:** Richard W. Brown **103R:** David Phillips **106TL:** John Colwell/Grant Heilman Photography **106TLC:** Grant Heilman Photography **106TCR:** Lynn M. Stone **106TR:** Lynn M. Stone **107(ALL)L:** Oxford Scientific Films/ANIMALS ANIMALS **108:** Brent Jones **109ALL:** Dr. E. R. Degginger **110CR:** Lynn M. Stone **112:** FPG **114:** Walter Chandoha **125:** David R. Frazier Photolibrary **127:** The Naval Research Laboratory **132:** David R. Frazier Photolibrary **134-135:** Norman Owen Tomalin/Bruce Coleman Inc. **138TR:** Arthus-Bertrand/Peter Arnold, Inc. **138BL:** Herman Kokojan/Black Star **138BR:** Dwight Kuhn/DRK Photo **143T:** Tim Bieber/The Image Bank **148TL:** R. Hamilton Smith **150:** George Hall/Woodfin Camp & Associates **152:** (c) Stephen Dalton/NHPA **155:** Milt & Joan Mann/Cameramann International, Ltd. **161T:** Milt & Joan Mann/Cameramann International, Ltd. **162T:** Milt & Joan Mann/Cameramann International, Ltd. **162B:** P. Vandermark/Stock Boston **168:** David R. Frazier Photolibrary **172:** NASA **174:** Harald Sund **176-177:** Tom Algire **177R:** Harald Sund **178T:** Larry Lee/West Light **178B:** Larry Lee/West Light **178TC:** J. Serrao **180-181:** Harald Sund **180L:** Lawrence Hudetz **183R:** Peter Menzel **186-187:** Ed Cooper **187T:** Milt & Joan Mann/Cameramann International, Ltd. **189:** Erich Hartmann/Magnum Photos **192TL:** FPG **192TC:** J.Cigano/FPG **192TR:** David R. Frazier Photolibrary **192BL:** J. Divine/FPG **192BR:** Joy Spurr/Bruce Coleman Inc. **194:** Kim Heacox/Woodfin Camp & Associates **196L:** Dan McCoy/Rainbow **196-197:** Lawrence Hudetz **198:** (c) 1987/Lawrence Hudetz **200R:** George Roth **201:** David Muench **202R:** R. Hamilton Smith **202L:** George Roth **203:** Lynn M. Stone **204:** Randy Taylor/Black Star **205:** Steven Gottlieb/FPG **207:** Lawrence Migdale/Science Source/Photo Researchers **210T:** D.P.Hershkowitz/Bruce Coleman Inc. **210B:** Andy Burridge/Bruce Coleman Inc. **212:** E. Nagele/FPG **214:** NASA **215:** NASA **216:** (c) 1986/John Apolinski **217:** Frank Whitney/The Image Bank **218:** NASA **219:** NASA **220:** NASA **221TL:** Luis Villota/The Stock Market **221TR:** Jules Bucher/Science Source/Photo Researchers **221BL:** Jeff Adams/The Stock Market **221BR:** Dennis Milon **224-225:** John Bova/Photo Researchers **225T:** John Bova/Photo Researchers **226:** David R. Frazier Photolibrary **230B:** Jerry Schad/Science Source/Photo Researchers **230TL:** Warren Faubel/Bruce Coleman Inc. **230TR:** Lenn Short/Bruce Coleman Inc. **232:** Marty Snyderman **237R:** Don and Pat Valenti **244:** Breck P. Kent **253R:** Carolina Biological Supply Company **253L:** Jim Markham/Bruce Coleman Inc. **258:** John Running/Stock Boston